AF709898

a summary of EARTH PROCESSES AND ENVIRONMENTS

DEPARTMENT OF NATURAL SCIENCES (SP)

John Tomikel, Ph.D.
Professor of Earth Science

ALLEGHENY PRESS
California, Pennsylvania—Pittsburgh, Pennsylvania
15419 1523

Library of Congress Catalog Number LC 74–28957
ISBN 0-910042-19-5 (cloth)
ISBN 0-910042-18-7 (paper)

Copyright © 1975 by Allegheny Press
Printed in the United States of America
All rights reserved

CONTENTS

ILLUSTRATIONS

PREFACE: A PERSONAL NOTE

Throughout my life I have been searching for an identity in the universe. I have tried to find and understand the meaning of life and the creator of life and the universe. My academic life began with a major study in literature and philosophy. I had hoped that these would lead me to my destination. These did not because scientific discovery made a shambles of literary speculation as well as current philosophical ideology. I then switched to the hard sciences of biology, physics, and chemistry. These helped me to understand the physical earth and gave me a greater awe of this creation. I was no closer to touching the creator or finding a meaning to life. Eventually I came upon the concept of earth science which was a merging of the sciences and the philosophies which I had previously studied.

The field of earth science which is the scientific study of the earth and its processes has brought me to an understanding that I am part of an energy system which is constantly changing. I am composed of elements which have sprung up from this energy system and I am presently a part of this system. Upon my death I shall return to some other part of this energy system. I take comfort in that knowledge.

As to the creator and the meaning of life, I still pursue these with as much vigor as ever. I hope that my mind is open to all argument. There has been some psychic phenomena in my life which I cannot explain but I do not speculate wildly upon these experiences. I feel that humans know almost nothing about life and this creation compared to what there is to know. I have written this book to summarize what I feel to be the important aspects of the earth processes and environments, not so much for you the reader's benefit but for my own. It is a catalog or home base from which I hope to venture out and return to add to its store. Perhaps a more thorough understanding of these will lead me to the creator and the meaning of life. Or perhaps these will point to some other direction in which I must travel.

John Tomikel
California, Pennsylvania
1975

1

SURVIVAL OF THE BIOSPHERE

There is a tendency for authors of articles and books about the earth to ignore the living organism which is neither air, rock, nor water. A summary of earth environments cannot be complete without mention of the association of the inorganic earth to the organic. The mass of living organisms on earth may be described as the biosphere. It includes the plants and animals which live in the land, in the air, and in the sea. It is almost a continuous blanket of plant and animal life moving very slowly from one environment to another. Life is an endless cycle of repetition with small variation from one generation to the next. Almost all animals, including humans, die within forty miles of the spot where they were born so we cannot think of animals and humans as a shifting entity. Humans remain in a circumscribed environment and very few change their habitat. When human life is considered in its proper environment it is best considered as a fixture within that environment. However, the agility and energy of humans can cause the tenuous quality of their environment to change enough so that it will be unhabitable to them. If humans should perish, the earth may continue to exist but what would be left would be an insensible world incapable of interpretation.

No one has satisfactorily explained and defined the meaning of life, but somehow it must be tied up with the well being of humans. Humans and their harmonic relationship to each other and their environment are certainly goals which must be adopted if mankind is to survive. Individual life is short but life must continue and we as individuals must be concerned about its continuation. In order for life to have meaning we must believe that the quality of life for humans is of paramount importance and that we should all work toward improvement of living conditions for all creatures.

Since this work is about earth processes and natural environments it does not deal with the catastrophic human problem of overpopulation and its social implications but this should undoubtedly be the primary concern of humans.

Where humans have sought more than the minimal needs of food, clothing, and shelter they have made the general environment unpleasant even though their immediate life may have been more comfortable. The atmosphere has become polluted with industrial wastes, the hydrosphere with industrial and domestic wastes, and the lithosphere is now slowly coming under abuse by an increasing population and its technology.

Atmospheric pollution comes in the form of industrial gases mostly the oxides, chlorides, and fluorides of carbon, nitrogen, hydrogen, and sulfur. These pollutants are further increased by automobile and aircraft emissions. The gasoline motor is a serious hazard to human life in populated areas and in highly industrialized unpopulated areas atmospheric pollutants tend to destroy vegetation and the life forms supported by that vegetation.

Pollution of lakes and the ocean is increasing at a rapid rate. For many years the ocean has been a dumping ground for trash and garbage of coastal cities. International controls will certainly be required as this dumping increases. Valuable fishing areas are slowly being destroyed since the disintegration and decomposition of these materials must rely on the gases and chemicals of the ocean itself. Changes in the oxygen ratio and changes in temperature cause changes in the organisms which are usually not of a beneficial nature for humans.

The lithosphere is increasingly being used for sewage and garbage disposal in inland areas. Underground disposal areas have been set aside for liquid wastes from atomic reactors, sewage accumulations, pulp wastes in paper production, acid mine drainage, and ore washings. These will eventually pollute the major underground water reservoirs, indeed if they are all not already polluted.

Regardless of human activities the earth processes will continue to operate. If humans build a pier out into the ocean it temporarily deflects the normal longshore flow of sand. Unless humans keep dredging the sand, nature will reclaim the coast. If humans drain the swampland they must keep pumping out water. If they fail to do this the water will rise once again and the landfill will eventually sink into the swamp. If humans build their homes in a river valley the water will periodically drive them out and reclaim its domain.

Humans are in a constant struggle with the earth processes

within their environment. In some instances humans have been able to utilize these processes to their advantage. Where humans have defied the natural processes they have been in constant struggle with them, expending energy to battle energy. Humans can survive best where they can live in harmony with earth processes. They can live in harmony with the earth if they understand its processes. It is hoped that this work will serve as a beginning to that understanding.

2

GENERAL EARTH FEATURES

Home is the earth and it does not seem likely that in the foreseeable future humans will be able to leave the earth and dwell elsewhere. Since the resources and the possibilities of the earth are limited it is of utmost importance that humans take stock of their environment and begin to live within the limitations of that environment.

To understand the earth environment we must understand the size and shape of the earth and its movements through space. These movements and the size of the earth give us our concepts of time, weight, direction, distance, day and night, seasons, and such phenomena as tides. Indeed the very existence of many earth features depends upon the relationships of the earth to its space neighbors.

The earth is shaped like a sphere. Close measurements show that it is lightly bulged in the northern hemisphere and slightly flattened at the poles. Its diameter through the equator is roughly 12,710 kilometers. Its pole to pole diameter is approximately 44 kilometers shorter due to polar flattening caused by the earths rotation.

The highest point on the earth is Mt. Everest in the Himalaya Mountain range of Asia. It is approximately 10,000 meters or ten kilometers high. The deepest part of the sea are the trenches of the western Pacific Ocean which are about 12,000 meters deep, or about 12 kilometers. Therefore the relief of the earth, that is, from its highest peak to its lowest depression is about 22 kilometers. When this 22 kilometers is compared to the 12,710 kilometer diameter it becomes rather insignificant.

To viewers from outer space the earth appears as a smooth surfaced sphere, the depressions and elevations hardly perceptible. The surface of an egg is much rougher than that of the earth by comparison. The pencil mark of a pencil drawn circle one meter is diameter is much thicker than the relief of the earth.

For many years scientists wondered about the shape of the earth and went to great lengths to prove that it was round. Such things as the shadow on the moon during an eclipse and circumnavigation were used as proofs. Stories, really legends, about the boy Columbus on the docks watching the ships come in gave the impression that he was the discoverer of the roundness of the earth. Actually the size and shape of the earth were measured about 200 B.C. by Eratosthenes. His trigonometric measurements were so accurate that only slight refinements were necessary almost two thousand years later.

The earth's gross features can be divided into three major categories based on the states of matter, these are gas, liquid, and solid. The gaseous part of the earth is the atmosphere which is an envelope of air surrounding the earth. It contains approximately 78 per cent nitrogen, 21 per cent oxygen with the remaining one per cent made up mostly of carbon dioxide, argon, and water vapor with smaller amounts of hydrogen, xenon, krypton, neon, and various industrial gases.

The liquid portion is water and is referred to as the hydrosphere. The largest areas of water are the oceans of which there are four, these are the Pacific, Atlantic, Indian, and Arctic with the largest being the Pacific covering some 164 million square kilometers. The oceans are salt water which is a solution of disassociated ions. The most abundant of the marine ions are those of chlorine and sodium. Other ions include those of magnesium, calcium, potassium, and sulfate. When sea water is evaporated the most common salt formed is sodium chloride followed next by magnesium chloride and sodium sulfate. If the earth were flat and the existing water were to cover it, the water would be about four kilometers deep.

The land and the rock mass of the earth is the lithosphere. Its surface area is about 146 million square kilometers. The largest subdivisions of the land mass are the continents which are Asia, Africa, North America, South America, Europe, Antarctica, and Australia. The largest of the continents is Asia containing some 43 million square kilometers.

If one ignores the separation of the land masses into continents and considers the world as composed of two large land masses and two smaller land masses he will have an easier time assessing its resources. It also becomes obvious that most of the land is

DEPARTMENT OF NATURAL SCIENCES (SP)

in the northern hemisphere and most of the water is in the southern hemisphere.

Earth Movements

The earth goes through five movements, two major and three minor. The major movements are rotation about its axis and revolution around the sun. The minor movements are wobbling as it moves through space, the slow shifting of the polar axis, and the movement of the earth along with the sun through the universe.

Rotation on the axis takes 23 hours and 56 minutes. It takes that time for one point on the earth to line itself up with one point on the center of the sun. We are accustomed to thinking that it takes 24 hours since that is the length of one day. This would be the case if the sun and the earth were not moving through space. Galileo had assumed the rotation of the earth by observing a swinging chandelier in the Sistine Chapel. He concluded that the shifting direction of swing of the chandelier was due to the earth rotating under it. Foucault later built a pendulum and proved the rotation of the earth. This coupled with the gyroscope which proved that a rotating body maintains its axis in one direction gave new impetus to the scientists who measured aspects of earth rotation. It is this eastward rotation which causes the moon and all celestial objects to appear to rise in the east and set in the west.

Rotation, of course, gives the earth day and night. If the earth did not rotate we would have six months of darkness and six months of daylight.

Revolution,or the journey around the sun takes 365-$^1/_4$ days. The earth sun distance varies with the average being about 149 million kilometers. The path of the earth, that is, its orbit, is elliptical or pear shaped. When the earth is closest to the sun, at perihelion, its speed is increased and when it is at aphelion, furthest from the sun, it slows down. This tendency of objects rotating around a central axis to increase in speed as it gets closer to the axis has many applications for earth phenomena. The average speed of revolution is about 25.6 million kilometers per day or 105,600 kilometers per hour.

The earth's axis is inclined 23-$^1/_2$ degrees toward the plane of its orbit which causes variations in the amount of daylight received at any place on the earth surface. Long days plus high sun cause "summer" and short days and low sun cause "winter." If the earth's axis is 23-1/2 degrees from the vertical it can be said to be 66-1/2 degrees from the horizontal. This is sometimes a better way to consider the axis tilt of the earth. This is especially true when one tries to visualize the earth spinning in space in reference to the sun.

Since the earth is tilted, the direct rays of the sun play upon the earth surface between 23-1/2 degrees south and 23-1/2 degrees north of its center. These northern and southern extents of the sun's direct rays are referred to as the tropics. Since the sun can only illuminate half of the earth at one time the tilt causes some places even though they should be in sunlight to be in shadow. When the southern pole of the earth is tilted away from the sun the direct rays are overhead at the northern tropic and the sun does not shine on the southern pole. Refer to figure 1. The southern extent of the sun's rays in this position extend to 66-1/2 degrees south which marks the southern circle or Antarctic Circle. It gets the name of circle since it marks the shadowline or the circle of illumination on the first day of winter in the southern hemisphere. Just the opposite occurs when the earth's southern pole is tilted toward the sun. The sun then is overhead at the southern tropic, the Tropic of Capricorn and its rays do not reach the northern pole and the circle of illumination only reaches 66-1/2 degrees north or to the Arctic Circle. These extreme limits of direct sunlight are the solstices. In the northern hemisphere when the sun is at its highest position over the northern tropic, the Tropic of Cancer, we refer to that time as the summer solstice. This occurs on approximately June 21 of each year and marks the first day of summer in the northern hemisphere. When the sun is at its lowest or southernmost position over the Tropic of Capricorn it is the winter solstice and marks the first day of winter, usually December 22 of each year. The foregoing discussion is based upon a position in the northern hemisphere.

When the sun is positioned over the equator or halfway between the poles it is said to be at the equinox position. The vernal equinox marks the entry of the sun's direct rays into the northern hemisphere and the beginning of spring, this occurs usually on March 21. When the sun is positioned over the equator on its

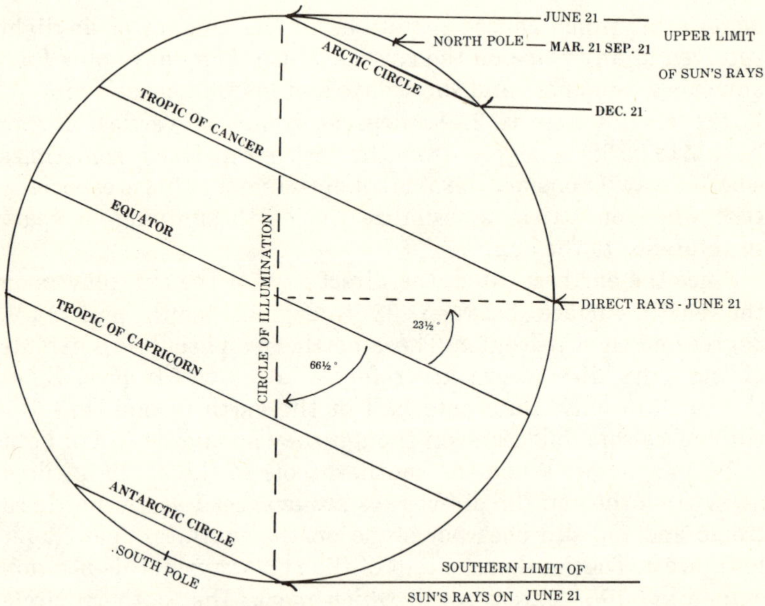

Figure 1: The sketch above illustrates points on the earth determined by sun position and the axis of rotation. The day indicated is the day of the summer solstice. The entire arctic region rotates under sunlight on that day while the entire antarctic region rotates in darkness. Where is the southern limit of the sun's rays on September 21? on December 21?

journey southward it marks the autumnal equinox and ushers in the season of autumn for the northern hemisphere. This usually occurs on September 21. Chart 1 summarizes this.

CHART 1

Approximate date	*first day of*	*position of sun's direct rays*	*event*
March 21	spring	equator	vernal equinox
June 21	summer	tropic of cancer	summer solstice
September 21	autumn	equator	autumnal equinox
December 22	winter	tropic of capricorn	winter solstice

If one considers the tilt of the earth to be 66-½ degrees from the horizontal plane he can visualize the reason for the land of the midnight sun. Refer to figure 1. On June 21 the entire area north of the Arctic Circle revolves in sunlight. The sunlit area dwindles slowly each day until at the equinox only half of the area is in sunlight and at the winter solstice all of it is in shadow. At the equinox the entire earth has 12 hours of daylight and 12 hours of darkness. The sun rises directly east and sets directly west for all places on earth on the day of the equinox.

The sun's direct rays cover 94 degrees of arc of the earth's surface in one year. The direct rays take six months to travel from one tropic to the other and six months to make the return journey.

One must bear in mind that the sun movement is apparent and it is the earth which is rotating and revolving. The term apparent sun is often used to keep this fact straight. The apparent path of the sun through the heavens is called the ecliptic. Surrounding the ecliptic, 15° on both sides, is the zodiac. When we say that the sun is in Pisces we mean that the earth has become aligned so that the sun is between it and the zodiac constellation of Pisces. The zodiac today differs in structure from the ancient ideas of the zodiac used by astrologers. Astrologers used the stars which make up the various figures of the constellations. Modern earth scientists consider the zodiac as consisting of 30° of arc on the heavenly sphere. The first day of a zodiac sign indicates the time that the sun will enter that designated portion of the sky.

3

EARTH'S INLAND WATERS

Water is second to air as the most important substance for human survival. This discussion will be confined to continental waters since marine waters are treated in another section. Man has two sources of fresh water available to him, that at the surface and that below the surface. Most of his domestic water supply comes from surface sources and only a small portion comes from underground.

Earth's water is constantly changing from solid ice to liquid water and then to gaseous water vapor and then back again. These changes of water are known as the hydrologic cycle. Basically it is thus. Water from the oceans, lakes, swamps, streams, and transpiration of plants evaporates and rises into the atmosphere. When it reaches a certain level the water vapor condenses to form suspended water which is visible in the form of clouds. The clouds move about and when they have reached a saturation level precipitation falls from them to the earth. This precipitation then returns to the atmosphere by way of evaporation and so on. Of the water which falls from the sky about half of it evaporates immediately, another third runs off into streams and about a sixth sinks into the ground.

Underground Water

Water that seeps into the ground is trapped in those materials which are porous, that is containing small holes or pockets. If the porous material is permeable, it allows the water to move through it then the water will continue downward until the air spaces or pores in rock or earth materials are saturated and the gases contained within those pores are forced to rise. Eventually a saturation zone will result, the top layer of which is the water table. The water table is not flat but takes the shape of the surface of the ground. From the water table to the surface the pore spaces are involved with exchanges of gases and liquids. Just above the water table is the capillary fringe, a belt where the water is working its way toward the surface by capillary action. Here water is being transferred upward by being passed from particle to particle much the way a towel absorbs water. This process is important in arid areas since it allows shallow rooted plants to benefit from the underground water table by transferring the water upward. At the soil level is the belt of soil moisture. Here water is retained in the pore spaces by plant root hairs and humus. The belt of soil moisture is the first stop for precipitation on its way to the water table. Between the belt of soil moisture and the capillary fringe is the intermediate belt where water is either moving downward or upward. In humid areas the general direction is down and in arid areas the direction is up.

Where the water table intersects the land surface a spring results. If the land is an enclosed depression a lake forms at the water table level.

A study of underground water indicates that the terms atmosphere, hydrosphere, and lithosphere are general since their boundaries cannot be well defined. In the ground, especially near the surface, there is rock, soils, water, and gas. It is a zone where the three general earth spheres mesh.

It is the underground water which is the source of supply for domestic wells and springs. The abundance of underground water is usually proportional to the precipitation of a region. In some instances the underground water may travel long distances in aquifers to areas of low precipitation. An aquifer is an underground structure which permits water to pass through it. Good aquifers are sand, gravel, and sandstone. Poor aquifers are clay, shale, and igneous rock. The movement of underground water depends upon the porosity and permeability of the aquifer. Usually water movement is slow in soils, only inches a day and in rock slower yet. Often people living in massive rock areas can improve their underground water supply by dynamiting the rock causing fractures to appear which will perform as aquifers.

The quality of ground water varies with the nature of the earth materials that it passes through and the degree of pollution by human activities. All ground water has dissolved minerals or salts in it. Some of these are beneficial but most are considered undesirable. The most abundant soluble salts in ground waters are from the compounds of calcium, magnesium, and sodium. Calcium and magnesium compounds produce hard waters; sodium compound waters are soft. Hardness of water is expressed in terms of parts of dissolved minerals per million parts of water. In limestone regions well waters often contain over 500 parts per million. Hard waters are bad for industry since these leave deposits of lime behind which cause boilers, pipes, and other containers to be constantly in need of cleaning and repair.

Lakes

Lakes are temporary features of the landscape and are all on their way to destruction either by drying up, filling with sediment, or by having their barriers eroded. Lakes are generally landlocked bodies of water so these would include the inland seas which are so named since they are salty. Salt lakes are created when evaporation of water excedes intake. These salty or saline lakes usually do not drain to the ocean. The lake environment is referred to as lacustrine.

Most lakes intersect the water table and obtain their waters from streams or underground drainage. Short lived or ephemeral lakes on deserts are called playa lakes, these are formed in low areas without drainage to the ocean.

There are many methods by which lakes may be formed. Glacial lakes may be created by glacier scouring, by till damming, or by ice dams. River associated lakes are created by shifting of meanders creating ox bow lakes or by changing the delta drainage and forming delta lakes. Karst lakes form by filling of sinkholes. Landslides create lakes by damming streams. Extinct volcanic craters which intersect the water table will form lakes. Block faulting may create lakes by creating rift valleys. If meteorite craters occur in humid regions then they will form lakes. Lakes may also be formed by damming through the efforts of man, plants, and animals. Beavers are notorious creators of lakes.

Lakes act as settling basins for streams. Silt laden streams will enter a lake and usually come out clean. It is this aspect of lakes which makes them highly susceptible to pollution since streams usually do not take with them what they bring to the lake. Pollution effects begin by raising the temperature of the lake thereby destroying life forms and if continued, pollution will eventually set up chemical combinations making life impossible.

Lakes have an effect on climate since they are a source of water vapor. Thus they increase precipitation. In cooler climates lakes may retard budding until the last danger of frost is past. Such is the case in the fruit belt which surrounds the Great Lakes and Finger Lakes. Lakes also regulate stream discharge reducing floods. They are of course reservoirs for domestic water supplies.

The largest lake in volume of water is Lake Baikal in the Soviet Union, it is also the deepest lake with its bottom some 1200 meters below sea level. The lake with the largest surface area is Lake Superior containing approximately 82,412 square kilometers. The Dead Sea has its surface 430 meters below sea level and is the lowest lake of the world. The highest lake of noteworthy size is Lake Titicaca of Peru which is about 4,160 meters above sea level.

Swamps

The swamp environment is referred to as paludal. Swamps are wet places created by their topographic association with water supplies and drainage. Thus swamps are usually low features with abundant vegetation. It has been estimated that two per cent of the world is covered by swamps.

Swamps may contain fresh or salt water and they may have varied conditions of origin. Swamps of the Great Lakes are created by undrained glacial conditions. The Everglade Swamp is an uplifted coastal plain. The lower Mississippi River area contains deltaic swamps. The middle Mississippi River area contains backswamps formed when natural levees prevent normal drainage into the river. The sea coast areas contain tidal marshes which are swamps created by inadequate or slow drainage once the high tides have shifted.

Mangrove swamps occur along sea coasts, such as that of Brazil. Some mangrove swamps exist along the Florida Keys. These feature the mangrove tree which spreads by touching its branches to the water and then sending down roots.

Muskeg swamps are large poorly drained areas of the north. They are rich in decayed vegetation and their soils are black in color. These are also forms of peat bogs which form from enclosed lakes growing over the moss and then eventually with

DEPARTMENT OF NATURAL SCIENCES (SP)

trees growing on the decayed moss areas. Further north of the muskeg area are the tundra swamps which are spongy masses of waterlogged soil and vegetation. Tundra swamps are usually frozen most of the year. They are in permafrost areas and the ground below them remains permanently frozen.

The large swamp areas of the United States are the Everglades of Florida, the Great Dismal Swamp of North Carolina and Virginia, the Okefenokee of Georgia, the Mississippi River swamps of Louisiana, and the many small undrained areas of the glaciated states of the north.

Using Inland Waters

In actual consumption the biggest use of inland waters is for domestic and industrial purposes. Most of the surface waters used for this purpose need treatment since they are likely to contain sediment and bacteria. Treatment of the waters include the addition of chemicals or aeration to destroy bacteria and the addition of coagulants or filtration to remove sediments and suspended materials.

A growing use of surface waters is that of hydroelectric power. Water for electric power in America has hardly been tapped since less than a third of the potential is used. The requirements for water power is a good source of water and a high fall. A small volume of water falling a great distance has the same effect as a large volume falling a short distance. Mountain areas are of course better for water power than valley or plains areas. In the United States the best areas are the Pacific coast, the Rocky Mountains, and the Southern Appalachian Mountains.

Of the world areas, Africa has the best water power potential of all the continents since it has large tropical rivers which fall great distances. In Europe the potential lies in the Alps as well as in Scandinavia; Asia has the Himalayas. In South America there are three districts with high water power potential, these are eastern Brazil, the Andes, and southern Chile. Australia is low and dry and thus has little water power potential, however its neighbor of 1500 kilometers, New Zealand, has adequate water power potential.

Building dams and catch basins for water power not only provides a source of water for domestic and industrial uses but also for irrigation. Dams also provide recreation as well as act to control floods.

One of the bigger uses of surface waters is that of transportation. River transportation is cheap but has the following disadvantages. The depth of the rivers fluctuates with the seasons. Young streams are interrupted by falls and rapids. Old streams shift their channels depositing silt and mud. High latitude rivers are covered by ice part of the year. Streams are not always the best route between consumer and producer. Also dock facilities are difficult to maintain due to variable river levels and river boats move slowly, especially upstream.

Present great fresh water routes are the Yangtze, the Zaire and the Amazon in other parts of the world. In the United States the Great Lakes is the world's greatest inland waterway. The Monongahela-Ohio-Mississippi River system is also well traveled and annually handles about 55 million tons of materials.

The recreational uses of fresh water cannot be discounted in an inventory of earth resources. Internationally famous are the lakes of Lock Ness in Scotland, Luzerne in Switzerland, and the millions of small lakes in Canada and Scandinavia. Governments are in a constant organization of recreational facilities connected with waters. The fastest growing industry in the country is that connected with camping. If this trend is to continue some government regulation concerning pollution will have to be enforced.

4

ROCKS AND THE LITHOSPHERE

In order to understand the earth and the processes which operate on earth some knowledge of the lithosphere is required. A knowledge of the rocks, the major constituent of the lithosphere, helps to explain the differences in the landforms produced by them, it explains the differences and qualities of soils and it indicates the limited uses which may be made of them.

Elements and Minerals

There are 92 naturally occurring elements on earth. Of these only eight make up more than 90 per cent of the earth. In order of abundance by weight these are oxygen, silicon, aluminum, iron, calcium, sodium, potassium, and magnesium. Oxygen makes up about 47 per cent and silicon 28 per cent of the earth by weight. By volume the same eight elements abound but with oxygen making up 93 per cent of the total and potassium, second with slightly less than 2 per cent of the total earth volume.

An element is a substance composed entirely of like atoms which cannot be reduced by ordinary chemical means. Each element has its own particular atoms which possess a set nuclear charge or number of protons in its nucleus. The proton count is then peculiar to only those atoms which make up the element.

Elements can combine to form compounds which can combine to form minerals. Minerals may be a single element or a combination of elements.

A mineral is defined as a naturally occurring homeogeneous substance formed by inorganic processes and having a characteristic set of physical properties, a composition and a molecular structure usually in crystalline form. This definition is a geologic definition and excludes organic compounds such as petroleum and natural gas which are economic minerals but not geologic minerals.

Minerals range in chemical composition from simple elements such as gold and carbon to complex chemical combinations such as the silicates which may be composed of as many as ten different elements. Minerals exhibit definite crystal forms as well as other properties such as hardness, tenacity, luster, specific gravity, transparency, cleavage, and fracture. These properties may be exact or within a limited range.

Minerals are usually classified according to their chemical structure. The larger groups are silicates, oxides, sulfides, carbonates, and sulfates.

The largest group of rock forming minerals, therefore the most abundant, are the silicates which have a basic structure of silicon and oxygen associated with other elements in minor amounts. The most important silicate minerals are olivine, augite, hornblende, and biotite mica which are called ferromagnesians and muscovite mica, orthoclase feldspar, plagioclase feldspar, and quartz which are the nonferromagnesians. Ferromagnesians are dark in color and heavy in weight while nonferromagnesians are light in color and light in weight. The ferromagnesians make up dense rocks of the earth's crust under the ocean while the nonferromagnesians dominate in continental surface rocks.

Oxide minerals compose a fairly large group. These are formed by the direct union of an element with oxygen. These include the iron oxides of hematite, limonite, magnetite, and taconite. Corundum is the oxide of aluminum, cassiterite is the oxide of tin, and ice is the oxide of hydrogen. Ice meets the geologic definition of a mineral and is included here to remind the reader of that definition.

Sulfide minerals are formed by the union of an element with sulfur. These include pyrite which is sulfur and iron, chalcocite with copper, galena with lead, and sphalerite with zinc.

The carbonates and sulfates are also abundant mineral categories. The most common of the carbonate minerals is calcite which is the major constituent of limestones. The most common of the sulfates are anhydrite which is calcium sulfate and gypsum which is the hydrated form of calcium sulfate.

Rocks

Minerals combine to form rocks. Rocks are generally classified into three groups called families. These families are igneous, sedimentary, and metamorphic.

Igneous rocks are those created from molten earth materials. If they are formed at the surface they are extrusive and if they are formed deep underground they are intrusive. All earth materials are used to form various kinds of igneous rocks, therefore they vary in their basic characteristics such as color, acidity, and weight.

The extrusive rocks are formed from volcanic eruptions or fissure eruptions which cool rapidly creating small crystals and a fine texture. These are lava formed rocks and include scoria which looks like burnt coal cinders, pumice a light frothy rock with many holes, tuff which is formed from compacted ash, and obsidian which is volcanic glass. Obsidian nodules are found in volcanic ash and form the gemstone Apache Tears.

More massive extrusive rocks are formed at the surface or very near to it. Due to this location they are cooled rapidly. Basalt is the most abundant of these extrusives; it forms many islands and plateaus. The Deccan Plateau of India and the Colorado and Columbia Plateaus of the United States contain much basalt. Basalt is a heavy dark fine grained rock composed mostly of ferromagnesian minerals. Rock which is composed of ferromagnesians is referred to as sima which is a contraction of silica and magnesium. Also prominent among extrusive rock is rhyolite which is light in color, light in weight, and fine grained. This rock is placed in the category of sial since it contains silicon and aluminum and is composed mainly of nonferromagnesians.

The intrusive rocks form deep underground which causes them to cool slowly and to have large crystals giving them a coarse texture. The best known of this type is granite which contains quartz and feldspars in large amounts and ferromagnesians in small amounts. Granites may be red, white, or gray in appearance. They are sialic rocks and are the coarse equivalents of rhyolites. The coarse equivalent of basalt is gabbro. Other intrusive rocks include syenite and diorite. A very coarse granite is pegmative which may contain crystals several centimeters in diameter.

The molten material from which intrusive rocks are formed is referred to as magma. The term igneous comes from the Latin "ignis" which means to ignite but there is no fire or oxidation connected with igneous activity so the process may be misnamed.

The materials which form sedimentary rocks are deposited by the surface agents of wind, ice, and water. Mostly they are acted upon by water which deposits them according to size and weight. Materials which are in solution are deposited according to their precipitation and solubility properties. Those formed from sediments are said to be clastic and are formed from the broken particles of other rocks. Thus shales are formed from mud and clay, sandstones from sand grains, and conglomerates and breccias from gravels and cobbles.

Loose sediments can become solid rock by several processes. Sediments may be compacted as water is driven off. Also they may be cemented by evaporation of the waters around the particles leaving behind a binding material.

The rocks formed from precipitation and crystallization from solution are considered nonclastic or chemical. Of these, limestone is the most abundant. Limestone may be precipitated from marine or fresh waters and takes many forms. Plants and animals also help to precipitate lime which eventually may go to form limestones. Chalk is of this nature. The composition of limestone is calcium carbonate and when small amounts of magnesium enter the structure it becomes dolostone. Rock salt, gypsum, and anhydrite are also chemical sedimentary rocks.

Coal is a sedimentary rock since it is formed from particles of materials which are piled up and acted upon by the forces of compaction, cementation, and chemical activity. Coal is organic in origin. In the formation of coal, plant materials are piled up, covered with other sediment and the major gases of the organic matter driven off and the carbon concentrated. The higher carbon content, the greater the hardness and desirability of the coal. In order to be coal however the rock must retain some of its original volatiles of nitrogen, hydrogen, and oxygen which aid in combustion.

In the discussion on igneous rocks it was mentioned that tuff was formed from volcanic ash. Tuff would be better classified as a sedimentary rock since the particles are sediments which await some lithification process to turn them into solid rock.

Metamorphic rocks are formed from the alteration of igneous, sedimentary, and other metamorphic rocks. Alteration may be due to pressure, heat, or chemical activity. Since these conditions exist deep within the earth it is safe to assume that metamorphic rocks are created there. The pressures are created by overriding materials and the bending forces of the earth. The heat is created by volcanism or radioactivity or friction. Chemical reactions are encouraged by heat as well as the passage of fluids and gases

DEPARTMENT OF NATURAL SCIENCES (SP)

within the earth through the rock structures. When the ions of gases are exchanged with the native rock as gases pass through them the metamorphism is known as metasomatism. The metamorphism created by large uplifts and pressures is regional metamorphism and the metamorphism created by contact with molten matter is contact metamorphism. Each type of metamorphism produces its own particular types and sequences of metamorphic rocks.

Sandstone produces a metamorphic rock known as quartzite and limestones produce marble. Marbles may be green, red, yellow, brown, as well as white. The differences in color of marble are due to the small amounts of impurities in the original limestone. Granites and other igneous rocks form gneiss and schists. A history of rock may be traced from mud which is

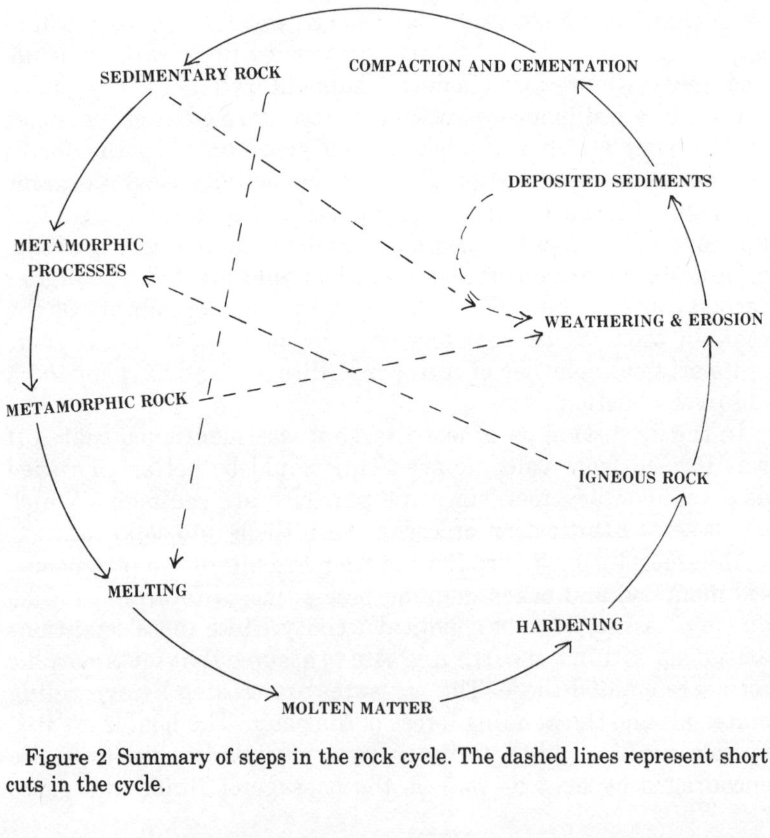

Figure 2 Summary of steps in the rock cycle. The dashed lines represent short cuts in the cycle.

compacted into the sedimentary rock shale. Shale is metamorphosed into slate, slate when acted upon by pressure becomes phyllite which in turn becomes schist. Schist is the most common metamorphic rock since it is an end stage in metamorphism. Schists are named for their most abundant mineral. Thus there are hornblende schists, biotite schists, and so forth.

The Rock Cycle

All things on earth seem to go through cycles, rocks are not different. The rock cycle is illustrated in figure 2.

If molten earth material crystallizes it forms igneous rock, igneous rock breaks off into particles and forms sediments, the sediments are acted upon to form sedimentary rock, these in turn are acted upon to form metamorphic rock which in turn is melted to form the magma and lava for igneous rock. Shortcuts in the cycle occur. Igneous rock may by-pass the sedimentary stage and go directly to metamorphism. Sedimentary rock may by-pass metamorphism and become melted. The cycle emphasizes the nature of earth materials and the fact that everything is acted upon by earth forces eventually returning to their original condition.

5

LANDFORMS

The earth's surface features can be divided into four general types of landforms; these are plains, hills, mountains, and plateaus. Some authorities may have variations but generally these designations prevail. For instance a valley can be considered as a definite type of landform, however since the four basic types contain valleys it is of no significance to have it as a division and only serves to complicate matters. Geologic definitions of the four landform types deal with underground structure and origin and are complicated. Relief definitions are much better. Under this system a plain is defined as an extensive area of level or gently undulating land of low altitude with a relief of 200 meters or less. Hills have relief of 200 to 700 meters and mountains have relief of over 700 meters. Plateaus are not always easily distinguished from other land divisions. They are characterized as broad uplands of high elevation possessing high relief.

Plains

Plains may be classified according to roughness, according to location, or according to origin. Each different classification system has its own uses in science. In the roughness category, plains with a relief of 20 meters or less are said to be flat, with a relief of from 20 to 50 meters they are undulating, with a relief of 50 to 100 meters they are rolling, and with a relief of 100 to 200 meters they are considered as rough dissected.

Plains classified according to location are generally broken down into coastal plains and interior plains. Plains classified according to origin are numerous since they may be created in many different ways. A stream eroded plain can be found in southern Ireland. Florida is a new coastal plain while the Paris Basin is an old coastal plain. The Sacramento Valley is an alluvial plain. A karst plain is created from underground water erosion and some of these may be found in Indiana in the United States and in Yugoslavia in Europe. Lacustrine plains are old lake beds, Mexico City is built on one of these. Loessial plains are created by wind deposition of dust, Northern Manchuria is of this type. Much of Hawaii is a lava plain. Canada and the Great Lakes region are glacial plains.

In North America the large plains areas are the Arctic tundra, the Atlantic and Gulf region and the Great Lakes. In South America the plains are along the Amazon and the Orinoco Rivers and the area of Argentina known as the pampas. Most of northwestern Europe stretching into Siberia are plains. Asian plains are along the great river flood areas including those of China and India. There are few plains areas in Australia and Africa.

The plains have long been the focal points of human activity. They are natural trading grounds since they connect the seaports with the hill areas. Transportation is easy on the plains and railroads and highways are inexpensive to construct. The bulk of the world's population lives on plains. In these areas the world's cereal crops are produced and the raw materials of the world are brought here to the cities for manufacturing. Of all landforms, man has made the most use of the plains.

Hills

Hills are rough areas that possess local relief between 200 and 700 meters. We must not be slaves to these limits since they are general. Sociologists can classify hills upon the basis of human activities. Some well known hill types are foothills which are found at the base of mountains, hogbacks with steeply dipping ridges, flatirons created by eroded canyons in arid regions, and piedmonts which occur between the plains and mountains.

Differences in hill lands are due to differences in bedrock structures, drainage patterns, climatic differences, their relationships to mountains and their past glacial history. The topography of hills gives them such names as ridges, folding hills, and badlands.

The biggest problem with hills is transportation. Early development of this country was hindered by the Appalachian Mountains which appear as low hills to the traveler today. Actually only a part of that structure is mountainous and most of it is hills. The early passes of historical significance through the Appalachians were the Potomac Valley route, the Cumberland Gap route, and the Mohawk Valley route. The searching for passes through hills has been repeated in all parts of the world in every age.

Hill people of course live in the valleys and their population patterns follow the drainage patterns of the large streams. Due to their isolation hill people in the past have developed a togetherness of spirit and in many cases their own speech habits. This isolation and different speech habit have earned them such names as hill billies, highlanders, and ridge runners. These nonstandard habits of customs and speech, together with clan loyalties have made hill people fierce fighters in the past and generally in battle the hill peoples have won out over the plains peoples.

Lumbering is the largest commercial industry of the hills with tourism fast overtaking it. Many hills have coal, oil, and gas deposits which makes this the major industry in some pockets of hills such as West Virginia and Kentucky. The biggest export of the hills however is people who migrate to the lowlands. The soil of the hills is usually poor except in small valleys and so the land is limited in its ability to support populations and for this reason the hill people of the Appalachians have drifted to the major cities found at the fringes of their domain, such cities as Detroit, Cleveland, and Chicago.

Major hill regions of the world include the Appalachians of North America, Wales and Scotland of the British Isles, the Carpathians of central Europe, and the foothills of almost all major mountain chains.

Plateaus

The definition of plateaus is not easy but generally they are considered to have high elevation, deep narrow valleys, and broad flat interfluves which are the areas between valleys. Their edges are often bounded by high steep ridges or cliffs. Plateaus are said to be tabular uplands having a relief of more than 200 meters.

The continent of Africa has plateaus composing almost three fourths of the land area which is the highest ratio for any continent. In North America there are the Colorado Plateau, Columbia Plateau, Allegheny Plateau, and Central Mexico. In Asia the Gobi Desert is a plateau as well as Arabia and Central India which is known as the Deccan Plateau. Europe has no large plateau regions except the central part of Spain and in Australia the great desert region is a plateau.

Plateaus classified according to location are intermontane plateaus found in high mountain areas, piedmont plateaus which are found between the mountains and the sea, and continental plateaus or tablelands such as those of Africa.

Plateaus classified according to origin are those of horizontal strata such as the Colorado Plateau, basaltic lava plateaus as exemplified by the Columbia Plateau in America and the Deccan Plateau in India, and uplifted erosional plateaus found in southern Brazil.

Plateaus in arid regions are characterized by blocklike divisions caused by stream erosion usually following fault lines. These beget names such as gulch, canyon, gorge, ravine, and escarpment. The Colorado Plateau has the Mogollon Rim and Hurricane Ledge marking its western boundaries.

Plateaus are almost impossible avenues for transportation. Airlines do a healthy business in plateau areas. In the middle latitude plateaus, sheep raising dominates the agriculture. In the tropics the plateaus have the most agreeable climates and are therefore the highest real estate values. Of world plateau cities there are Mexico City in Mexico, Madrid in Spain, Addis Ababa in Ethiopia, Sao Paulo in Brazil, La Paz in Bolivia, and Lhasa of Tibet. South American plateaus were the home of the early civilizations since these afforded protection as well as a pleasant climate.

Mountains

Mountains are distinguished by the smallness of their summit areas, their large steep slopes, and their high relief. Generally the slopes do not extend very far in one direction and they twist and turn as they connect to the lowlands. Since mountains have small summit areas terms such as crest, peak, horn, pinnacle, and mount are used and give a visual impression of the condition of those features.

DEPARTMENT OF NATURAL SCIENCES (SP)

Mountains can be classified according to relief just as any other landform. Those with relief between 700 and 1000 meters are low, between 1000 and 1500 meters are rough, and over 1500 meters are rugged. The term sierran is sometimes used to indicate any mountain with over 2000 meters of relief.

The major mountain range of North America is the Rockies. The Appalachians are only partially mountainous. The same situation is true of the Adirondacks and the White Mountains. In the far west the Cascades and the Sierra Nevada compete with the Rockies in grandure.

South American mountains include the Andes, the Guiana Highlands, and parts of eastern Brazil. African mountains are in small chains of the Atlas mountains in the northwest and the mountains of Ethiopia in the northeast with Mt. Kilamonjaro a famous landmark in the central portion. The Himalayas of Asia are the highest mountains of the world. Other Asian mountains are found in Turkey, Pakistan, and Manchuria. The famous Urals are really best classified as hills. The European mountains include the Alps of Switzerland, the Pyrenees between France and Spain, the Appenines of Italy and the ranges of Norway. There are numerous island mountains of which New Zealand is the most densely populated.

Most classifications of mountains deal with origin. Mt. Fujiyama is a volcanic mountain, the Black Hills of South Dakota are dome mountains, the Middle Rockies are laccoliths, the Sierra Nevada are block mountains, the Appalachians are folded, the Northern Rockies are overthrust fault mountains, the Great Smokies are classified as complex since they are folded and faulted, and the Catskills and White Mountains are erosional. There are no glacial mountains since glaciers merely modify existing mountains.

Mountains have always been areas of sparse population and have been a handicap to travel and communication. They also affect weather conditions with their windward slopes being wet and their leeward sides approaching desert conditions. Passes through large mountain chains are famous in history as well as in present day communication. Famous passes include the Brenner and St. Gothard of the Alps, the Donner and Cajon of the Sierra Nevada, and the Khyber between Pakistan and Afghanistan.

Major economic activities of the mountains include lumbering, mining, and tourism. Most farmers living in mountainous regions practice transhumance, that is, they migrate up and down the slopes with the seasons to take advantage of the upper meadows when the snow has melted.

There is a zonation of climates on most mountains and a traveler going from the base of the mountain to the top will encounter different climates up the slopes in the same order as if he were traveling from the mountain base to the poles. At the equator one traveling up the mountain would first encounter the tropical crops of bananas and cacao. Further up would come coffee and then grains and potatoes. Further climbing would introduce deciduous forests which change to evergreen and then into meadows and finally permanent snow.

People who live in mountains live in the valleys of mountains and except for tourism have small contact with the rest of the world. To an outside observor they live lives of solitude in beautiful surroundings. This solitude however is occasionally interrupted by avalanches and landslides which are the biggest hazards to mountain living.

Landforms Within The Ocean

If the oceans were suddenly drained of water forms similar to those of the continents could be observed. These ocean forms have special names. The area where the continents and oceans meet is the littoral or tidal zone. During the period of one day it is alternately covered with water and exposed to air. From this tide zone the sea floor slopes gradually on an almost flat plain called the continental shelf. It averages about 70 kilometers in width and extends to a depth of about 200 meters. It is the neritic zone. From the shelf a sharper but still gradual dip occurs, this is the continental slope or bathyal zone. It slants about 10 degrees and extends from 200 to about 2000 meters below the water surface. Below the bathyal zone is the ocean deep or abyssal zone which comprises the major part of the ocean bottom. It is about 3000 meters at its deepest and its bottom is often referred to as the abyssal plain.

Beside these general ocean zones there are some peculiar features which exist. In the middle of most oceans there are ridges. The Mid-Atlantic Ridge extends the entire length of the

Atlantic Ocean. Oceanic ridges rise to 4000 meters above the sea floor. Iceland is a part of this system. In the Indian Ocean ridges extend from India to Antarctica. Ridges are the result of volcanic activity created when the sea floor spreads as continents drift from their original positions.

Guyots and sea mounts are also mountains beneath the ocean. The guyots are flat topped and the sea mounts are peaked, both are isolated mountains which are abundant from Alaska to the middle Pacific Ocean.

The deepest part of the oceans are the trenches which are curved and associated with island arcs. The deepest of these occur in the Pacific, here they reach depths of 12,000 meters. Their association with islands suggest a common tectonic origin.

Another low feature of the ocean is the submarine canyon which are deep linear ravines off shore. These occur off the coasts of some large rivers such as the Hudson River of North America and the Zaire River of Africa. Many theories have been advanced to explain their origin. Two which are particularly appealing are that the canyons were created by river sediments scouring them out and or they were created by the rivers which emptied into the sea much further than their present location, this occurred when much of the ocean water was tied up on land in the form of ice during peak glacial periods.

Landforms of Arid Regions

Arid regions are changed mostly by physical weathering and removal of these weathered materials under the influence of gravity. A typical mature arid region landscape consists of a mountain range with a gentle slope called a pediment at its side. The pediment leads down to a low basin known as a playa. The pediment is cut by the major forces at work in the arid region which is mainly water erosion with some sand blasting by wind. Weathering occurs upon the pediment and when it does rain this weathered material is quickly washed into the lowland playa.

Flat topped hills in arid regions are mesas and buttes. These are created when a resistant caprock, usually sandstone, protects weaker rocks underneath, usually shales. As the caprocks break the hill recedes since the shales do not hold up well.

The bases of the cliffs in arid areas are surrounded by talus. This is rock which has broken from the cliff and has fallen to the cliff base. The talus forms a slope leading away from the cliff. Where water is involved, mostly in mountain regions, it deposits material in delta fashion upon the land. These are known as alluvial fans and these may spread out for many kilometers at the base of the mountains and may be many hundreds of meters thick.

6

ELEMENTS OF THE ATMOSPHERE

The atmosphere is a mechanical mixture of several gases which varies slightly in composition with latitude and weather conditions. The major atmospheric gases are nitrogen and oxygen with nitrogen making up approximately 78 per cent and oxygen 21 per cent of the mixture. The remainder consists mostly of water vapor, carbon dioxide, and argon with minor amounts of hydrogen, helium, neon, krypton, and xenon.

The atmosphere may be divided into three zones based on physical conditions. These are the lower zone or troposphere, the middle zone or stratosphere, and the upper zone or ionosphere.

Troposphere

The troposphere extends from the ground surface to a height of about 15,000 meters. The top of the troposphere is marked by the tropopause which can be defined as a physical region in which there is a change in the decreasing trend of temperature. In the troposphere the temperature of the air decreases with altitude at a steady rate. It is this decrease in temperature which causes moisture laden air to condense and produce clouds as the air rises.

The troposphere contains most of the atmosphere by weight. It is in this zone that most of the observable weather takes place. Here the moisture of the atmosphere is contained and it is shifted about by winds and other mechanical transformations.

Stratosphere

The middle zone or stratosphere extends from 15,000 meters upwards to an altitude of about 80 kilometers. It is a region of horizontal air motion. It is here that the high winds known as the jet stream are found. Sometimes these reach velocities of 300 kilometers per hour. Generally the jet stream moves from east to west.

The stratosphere is usually devoid of clouds but occasionally some of the higher varieties are found in the lower stratosphere. Stratosphere temperatures are more or less constant at about −50° to −60° Celsius. In some areas of the stratosphere chemical changes and concentrations of ozone may cause temperatures to rise as high as 65° C. These warm zones are known as the chemosphere, ozonosphere, and mesosphere depending upon their origin and content. Man cannot live in the stratosphere without temperature, pressure, and oxygen control. The upper layer of the stratosphere is called the stratopause.

Ionosphere

The upper zone of the atmosphere is the ionosphere. This is a region of slightly increasing temperature. This layer is composed of electrified particles or ions. The ions are caused mostly by solar radiation acting upon the gases giving them an extra electron making them electrically negative or taking one away making them electrically positive.

The ionosphere extends to the limit of the gases surrounding the earth. It is divided into several layers, each division based upon physical properties. The layers are labeled D, E_1, E_2, F_1, F_2, and G. These different layers serve as a ceiling upon which radio and television waves are bounced back to earth and it is this bouncing which enables people to receive radio and television broadcasts. Since radio and television waves travel in straight lines they must be reflected in order to be received around the curvature of the earth, otherwise the curvature would inhibit reception beyond a range of approximately 40 kilometers.

The D and E layers of the ionosphere exist only in the daytime as long as the sun is shining on them. The E layer reflects the AM band on radio, the F layers the AM and FM bands as well as TV. The F layers combine into one layer at night.

When the sun goes down it usually still shines upon part of the ionosphere. At certain times of the year these ionized gases are charged to the point of glowing. In the northern hemisphere these northern lights are the aurora borealis and in the southern hemisphere this glow is the aurora australis.

Some authorities on space and the atmosphere maintain that it is best to consider the ionosphere as extending to the limit of the F layer and that the outer portion of the fine gases be called the exosphere. This makes a classification more convenient and allows the incorporation of the Van Allen belts into the exosphere. These are belts of intense radiation over 1500 kilometers above the earth. There is at this time no measurable connection between these belts and terrestrial earth phenomena.

WEATHER ELEMENTS

When we hear the news broadcasts and the weather reports are given, there are usually five factors covered. These are temperature, pressure, moisture, clouds, and winds. These may be considered as the basic elements of weather. Weather is defined as the condition of the atmosphere at any given time for any given place.

Insolation and Temperature

Energy received at the earth's surface from the sun is called insolation. Of the solar energy emitted, the earth only receives about one two billionth of it. Of this amount about 40 per cent is reflected back into space, another 20 per cent is absorbed by the atmosphere, and 20 per cent more is diffused or bounced around in the atmosphere leaving only 20 per cent to reach the earth surface. The amount of solar energy received at any given spot on the earth depends upon the distance of the earth from the sun, the angle of the sun's rays, the length of daylight, fluctuations of solar radiation and the condition of the atmosphere as well as the surface of the earth.

Insolation has the greatest noticeable effect upon land rather than water since land heats up five times faster than water. It also cools five times faster. Therefore the greatest contrasts of temperatures are found in the interior of continents where summers and winters are more extreme than near the sea coasts. Large lakes also have an effect upon local climates causing them to be milder than they ordinarily would be.

Insolation is of major importance in controlling the life of humans and their environment. It influences their types of homes, it influences the development of soils, it limits the plant types of an area, it determines the length of the growing season, and it determines the type of clothing humans wear.

One facet of insolation is the concept of temperature. Everyone knows that a thermometer measures temperature but if you asked someone to define temperature, the chances are they could not do it. It is difficult and a definition is hard to find. Here is what happens. Atoms have a nucleus composed of neutrons and protons. Around this central nucleus are planetary electrons in orbits or shells. As heat is added to the atmosphere the electrons of the atoms which compose the molecules of gas move further out in the shells, this causes the molecules to move faster since they are lighter. As their speed increases there is more bombardment of the thermometer by the molecules of gas. More bombardment causes more pressure and the liquid in the thermometer rises. As heat is removed, the molecules slow down and we say that the temperature is dropping. Therefore a thermometer measures the activity of molecules and temperature is an index of the activity of molecules.

When temperatures are plotted on a map they are usually given for an average time such as one month. Lines are sometimes placed on maps connecting areas of equal temperature, these are isotherms. The two months when the average isotherm patterns show significant fluctuation are in the coldest and warmest months of January and July. In January the isotherms over land in the northern hemisphere bend toward the equator. This is due to the unequal heating effects of water and land. What this indicates is that one would have to go further south in the middle of a northern continent in winter to get the same temperature enjoyed by a more northern sea coast location. For these same reasons the isotherms over land bend toward the poles in summer. Isothermal patterns are not only influenced by land and water bodies but by wind directions, ocean currents, and differences of moisture in the atmosphere.

Pressure

The atmosphere has weight. Atmospheric pressure is the weight of the atmosphere at any given point of measurement. The pressure exerted by the atmosphere is the same at any given point and in all directions. The weight varies as one moves any appreciable distance from one spot to another. Air moves from areas of high pressure to areas of low pressure; therefore, knowing the pressure of the atmosphere is a convenience for predicting which way the weather will move.

Standard air pressure is based on an agreement reached by meteorologists to facilitate weather prediction. It is defined as the sea level pressure halfway from the equator to the poles at a temperature of 0° C. It is important to state the temperature since the weight of air varies with temperature, warm air being lighter than cold air. Pressure can be stated in many ways depending upon which is most convenient to the user. Standard air pressure is given as a comparison force. It is often compared to the weight of a column of mercury and in this instance it is equal to the weight of a column of mercury 76 centimeters high. The early barometers which measured the weight of the atmosphere used water for comparison and in this instance it was equal to the weight of a column of water over ten meters high.

For accurate weather prediction millibars are used. These take into consideration not only the weight of the atmosphere but the rate at which it is being pulled toward the center of the earth. In this system, standard air pressure is 1013.2 millibars.

Atmospheric pressure is measured with a barometer. There are two principal types. The mercury barometer is a column of mercury encased in glass. It stands an open end in a pool of mercury and has a vacuum at the top of its closed end. As the air pushes down on the open pool of mercury it keeps the mercury in the tube from falling by its weight, that is the weight of the atmosphere. The weight of the standing mercury in the tube is equal to the weight of the atmosphere.

Another barometer is the aneroid barometer. It is more convenient and does not break as easily as the mercury barometer. It contains hollow metal cells about four centimeters in diameter. These are vacuum cells and are connected in series and eventually connected to a needle which points to a dial. Atmospheric pressure acts on the cells and causes the needle to move thus indicating changes in pressure.

Barometric pressure can be automatically recorded by the barograph. Instead of a pointer needle connected to the pressure sensitive cells there is a pen. A cylinder containing graph paper rotates on a spool much like a clock. As the cylinder moves, the pen constantly records the pointer position which is a record of the atmospheric pressure.

Isobars are lines on a weather map connecting areas of equal pressure. Thus high and low pressure areas can be identified. These not only can be used to predict the direction of air movement but also to identify the direction of winds which would be associated with them. Winds in the northern hemisphere move counterclockwise around a low pressure area and are referred to as cyclones. Also the winds move clockwise around a high pressure area and are called anticyclones.

Air pressure becomes lower with increasing elevation since less air remains above the higher elevated points. Regardless of actual readings in an area, all pressures are reduced to what they would be at sea level for purposes of computation and prediction.

Pressure varies with changes in temperature with warmer air reducing pressure since warm air is lighter than cold air. Water vapor also makes air lighter since its molecules weigh less than those of nitrogen and oxygen which it replaces in a volume of air. Thus warm moist air is lighter than cool dry air.

Humidity

Water vapor exists in the air in most areas of the world. Even the driest desert usually has some water vapor in its atmosphere. Water in the atmosphere exists in both visible and invisible forms. The visible forms are fog and clouds. The invisible forms make us feel hot and sweaty on warm days. It can become visible when a cold pitcher of liquid reduces the temperature around it and water vapor condenses upon the container.

Water vapor keeps the temperatures of the earth from reaching extremes of hot and cold. As water changes from gas to liquid to solid it gives up heat. As it changes from solid to liquid to gas it absorbs or takes heat. Thus the changes of state of water keeps heat in the atmosphere by constantly utilizing it.

Water vapor in the atmosphere is referred to as humidity. Absolute humidity is the actual amount of water vapor in a

DEPARTMENT OF NATURAL SCIENCES (SP)

given amount or volume of air, this is measured in grains. Relative humidity is the amount of water vapor in the air compared to the amount that it could hold at that temperature. Changes in air temperature cause the air to change the amount of water vapor it can hold before condensing. The dew point is the temperature at which water vapor condenses. Thus when the night air temperature is lowered we find dew on the lawn in the morning. Or when the air is heated at the surface and it rises to a few thousand feet it experiences a drop in temperature and clouds are formed.

Humidity of the atmosphere is measured by a hygrometer which has two thermometers, a dry bulb thermometer which is bare and a wet bulb thermometer which is covered with cloth at its bulb end. The wet bulb is moistened by immersing the cloth covering in water. What is measured is the drop in temperature on the wet bulb which is a measurement of the rate of evaporation. By calculations comparing wet and dry bulb readings the relative humidity can be determined.

A sling psychrometer is a variation of the hygrometer. With this instrument the thermometers are swung through the air thus causing the evaporation process of the wet bulb to be more rapid.

A hygrograph is an instrument which has a pen on one end of a small pointer rod which records on graph paper. The other end is tied down with human hair and is made tight by a pivot in the middle of the rod. The hair reacts to changes in humidity and the pen is moved at the other end which causes it to make marks on the graph paper which is placed on a revolving drum which is run by a clock mechanism. Thus the hygrograph records the humidity directly. Anyone with long hair knows that hair reacts to humidity in the atmosphere. The Apaches knew this when they used the phrase "when the locks hang limp in the scalp house surely it will rain."

Clouds

Clouds are visible water vapor which has condensed in the upper atmosphere. The particles are small and only when the condensed particles coalesce enough to become heavier than air does it rain.

There are two basic types of clouds, the stratus which are clouds in layers and the cumulus which are fluffy clouds. All other clouds may be considered as variations of the two basic types.

Clouds can be classified according to altitude. Thus there are high, middle, and low clouds. The high clouds are found from 6,000 to 15,000 meters. The three major high clouds are cirrus which are little wisps of clouds, cirrocumulus which are extremely high small fluffy clouds, and cirrostratus which are very high clouds in layers. Cirrostratus clouds are responsible for creating rings around the moon. They are the forerunners of rain and thick cirrostratus clouds usually indicate rain in about two to three days.

Middle clouds are found from 2,000 to 6,000 meters. The two basic types are altocumulus and altostratus; these need no further explanation except to say that they are thicker than their cirrus counterparts.

Low clouds are the stratus and cumulus. Sometimes the cumulus are so close together they are called stratocumulus. When cumulus clouds build up on hot summer days they are referred to as towering cumulus. When these build up rapidly they create clouds known as thunderheads or anvilheads which are cumulonimbus clouds. These produce rain, hail, and lightning. They cover only a few kilometers in diameter and move very rapidly. The rain beats down for about twenty minutes as they pass over. Usually cumulonimbus clouds travel in sets and so the rain is intermittent but in heavy downpours. If rain is produced from stratus type clouds they are termed nimbostratus. The term nimbo attached to clouds means rain. The rains from nimbostratus clouds are more gentle and enduring than rain from cumulonimbus.

Clouds are formed when water vapor condenses by the cooling of the air. The small invisible particles that the water vapor condenses upon are condensation nuclei. These are usually minute particles of dust inland and minute particles of salt in sea areas. If it were not for the nuclei there would be no condensation. For instance, if you breathe out rapidly in front of you, nothing happens but if you go to a window and breathe upon it the chances are moisture will collect on the window, the window acts as a condensation surface. What you do when you breathe on the window is to add more moisture to a small area and give a surface on which it can collect. So clouds and fog, which is really a cloud at ground level, can be formed by cooling the atmosphere or by adding more moisture to it and providing condensation nuclei. There is always some condensation nuclei in the atmosphere since the atmosphere is never really free from impurities.

Rain can be created by saturating a cloud with condensation nuclei. This gives more surfaces for vapor to condense upon and when these particles become heavy with condensation it rains. Nuclei used for this purpose are usually dry ice or silver iodide which has the same crystal structure as dry ice. Most states have laws concerning rain making so this must be done with strict government supervision.

7

THE DYNAMIC ATMOSPHERE

The vast amounts of energy contained in atmospheric processes are most evident in the movement of air. Wind is simply moving air. We do not see wind but things which are being moved by the wind such as flags, kites, and trees. We feel wind putting pressure upon us. If the land surface were not rough and the water surfaces evenly heated the winds on the earth would be easy to predict, but since these conditions do not exist the winds are varied.

Winds are named for the direction from which they blow. There are cold north winds and warm south winds in the northern hemisphere.

Zonal winds are created by unequal heating of the land and thus the uneven heating of the air above the land. At the equator the air rises since it is heated rapidly. It rises and spreads north and south. Refer to figure 3. At the surface air rushes in to the equatorial area to make up for the air which is rising, this causes air to descend at the tropics. The desending air at the tropics spreads north and south. The air heading north in the northern hemisphere rises around the Arctic Circle since it is pushed above the cold air coming from the pole. Thus there are three cells of atmospheric circulation. One cell operates from the equator to the tropics, one from the tropics to the circles, and one from the circles to the poles. At the earth's surface the air moves from the tropics in two directions, that is, to the equator and to the circle. It moves from the poles to the circles.

In the northern hemisphere the winds are deflected to the right because of the rotation of the earth. This phenomena is called the Coriolis Effect. In the southern hemisphere winds are deflected to the left. Thus the air moving toward the equator comes from the northeast in the northern hemisphere and from the southeast in the southern hemisphere. These are the trade winds and are labeled northeast trade winds and southeast trade winds respectively. The area of rising air over the equator produces the feeling of no wind or calm since one cannot feel

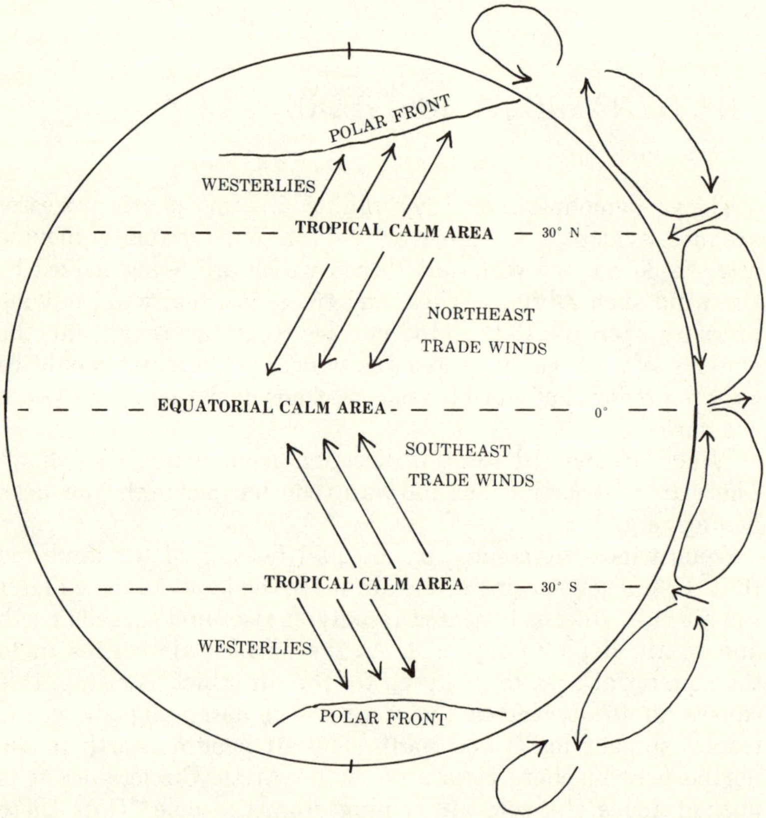

Figure 3. The sketch illustrates normal air movements on the rotating earth. Air rises over the equatorial region and descends over the tropics. This is illustrated to the right of the sphere. The arrows on the sphere depict wind movement along the earth's surface. The zones above migrate slightly with the sun position and the seasons. Due to rotation the winds in the northern hemisphere are deflected to the right and the winds in the southern hemisphere are deflected to the left.

a breeze from rising air. This area is warm and humid and is known as the doldrums or area of equatorial calm.

The air moving from the tropics to the circle is deflected to the right in the northern hemisphere and to the left in the southern hemisphere giving both a westerly origin and so they are called westerlies. The areas over the tropics, the areas of descending air, are also an area of calm each known as the area of tropical calm. These are drying air movements since descending air absorbs moisture. The tropical deserts are located in these areas of tropical calm. The deserts of Sonora in Mexico, Sahara in North Africa, Arabian in East Africa, Kalahari in South Africa, Atacama in South America, and Victorian in Australia are all in the area of tropical calm. This area is also called the horse latitudes. It gets this name from the early sail using explorers who were trapped in the calm on their way to the new world. Some were forced to eat their horses when they ran out of other food. Some are reported to have thrown their horses overboard to make the ship lighter for rowing.

Polar winds move from the northeast in the northern polar region and from the southeast in the southern polar area. These are also affected by high pressure systems and so their movements are not always predictable.

There are also local winds which are usually caused by unequal heating of water bodies. During the day in sea coast areas the warm air rises over the land since land heats faster than the water. Air moves in from over the water since sea air does not heat as fast as land air, thus a sea breeze is created. At night the reverse is true, the sea is warm and the land cools rapidly. Air rises over the sea and is replaced by air from over the land thus the night breeze is a land breeze.

In mountain areas there are breezes which are created by descending air. The air at the top of mountains, especially if they are covered with snow is heavier than the air at the base of mountains. This air is further dried by being drained of its moisture on the side of the hill which faces the prevailing winds. The windward side of hills is usually moist since the air is forced up to colder elevations where condensation and precipitation occur. The lee side then gets air which is descending and drying. Thus arid regions are found on lee sides of mountains. These descending drying winds are known as chinooks, foehns, and siroccos.

Air Masses

An air mass is a large portion of the atmosphere that has uniform properties throughout its volume. These contain nearly constant horizontal values for temperature, humidity, and pressure. An air mass develops if the air of a region remains at rest or stable for a period of time over a uniform surface. Such surfaces are low land areas and areas of the sea.

Once an air mass has developed high pressure it moves out of its source region to another location, usually toward low pressure areas. Air masses are labeled according to their source region, thus there are continental and maritime air masses and there are polar and tropical air masses. These two designations are the most common. Letter symbols are used to designate their location on weather maps. Thus a cP air mass is continental and Polar in origin. Such air masses originate over Siberia and Canada. Or an air mass may be mT, that is, marine Tropical. These form over the Caribbean and the Gulf of Mexico. People living in the middle latitudes are constantly in changing weather due to cold air masses moving from the poles and warm air masses moving from the tropics. Extremely cold air masses are labeled A for Arctic or Antarctic and extremely hot air masses are labeled E for Equatorial. A large S denotes a superior air mass, one which is the result of subsidence from a higher altitude. Air masses may also be labeled k for colder than the surface over which it moves or w for warmer.

Fronts

The leading edges of air masses are fronts, warm air masses have warm fronts and cold air masses have cold fronts. Since warm air rises, the shape or slope of the warm front is at an angle. Refer to figure 4a. The leading portion is high in the sky and tapers down to ground level where the warm air is sufficiently strong to push out the existing air of an area. The passage of a warm front is marked by slight wind shifts, distinct rises in temperature, slow clearing of clouds, rapid increases in humidity, and in the northern latitudes a wind shift to the south. Rain associated with a warm front is usually from nimbostratus clouds and precedes the arrival of the front at the ground surface.

Cold fronts have a rounded shape with the leading edge at the ground. Refer to figure 4b. Since it is colder than the air of an area it moves under it. The passage of a cold front is marked by a rapid drop in temperature, a great shift in wind

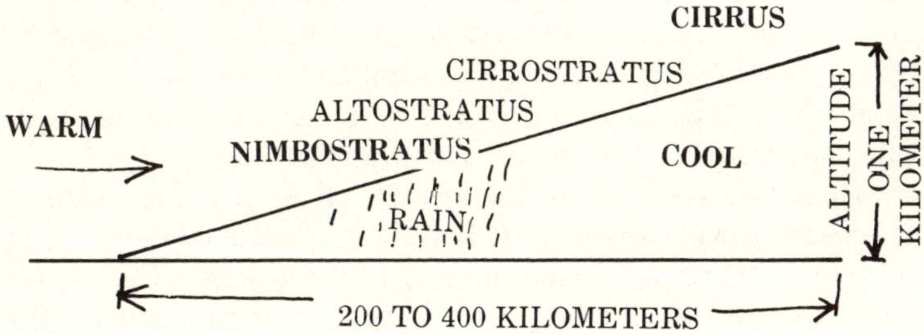

4a

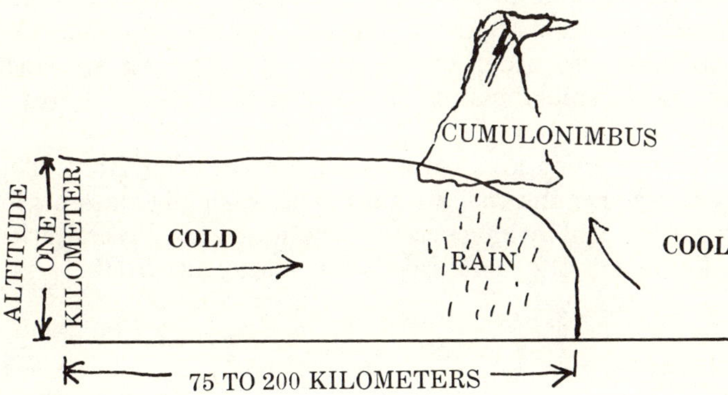

4b

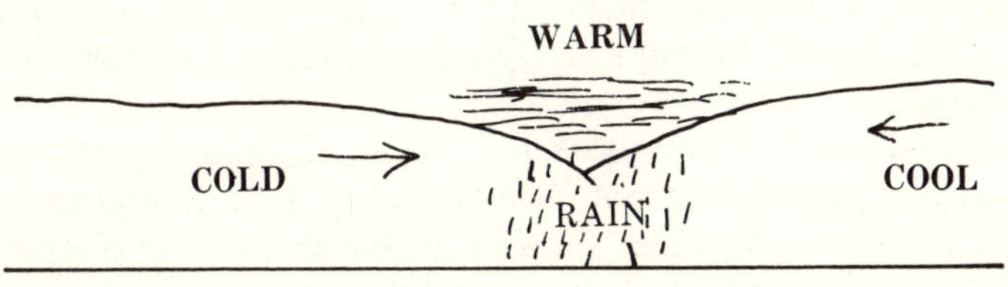

4c

Figure 4: Cross sections of major fronts. 4a—warm front 4b—cold front 4c—occluded front

direction, a rapid increase in atmospheric pressure, a rapid clearing of clouds, a marked decrease in humidity, and winds in the northern hemisphere coming from the north, northeast, or northwest. Rain from a cold front is usually from cumulonimbus clouds and is located behind the front edge.

When two cool fronts move into an area at the same time the warm air is shoved aloft and a temperature inversion exists. A temperature inversion is a condition when the upper air temperature is warmer than the ground temperature which is not normal since it should be colder with altitude. Such a frontal condition is an occluded front. Refer to figure 4c. This is a situation when air pollution is most evident since the warm air aloft acts as a ceiling and does not permit rising smoke and fumes to penetrate and dissipate in the upper atmosphere. Temperature inversions can also occur in the evening with still air when the land cools rapidly making the surface air cooler than the air aloft.

A stationary front is a variation of an occluded front. This occurs when warm and cool air meet. The warm air mass begins to override the cold air mass cooling the warm air. Precipitation lasting for days is the usual result of a stationary front.

Storm Types

Most of the middle latitude areas are affected by cyclonic storms. This is the constant rhythm of cold and warm air masses passing over. These are the result of high and low pressure systems moving across the middle latitudes in harmony with the natural westerly flow of air. The winds associated with cyclones and anticyclones usually last for days and blanket a large area. Associated winds may be severe or gentle depending upon the differences in pressure. The term cyclone should not be confused with Tropical Cyclone which is another term for hurricane.

Thunderstorms are those storms which produce thunder and lightning. These are not designated as thunderstorms by the United States Weather Bureau unless the thunder is actually heard at the weather station. The phenomena of heat lightning is actually a thunderstorm in another area, the observor just doesn't hear the thunder but he sees the atmospheric light display.

There are many differences between thunderstorms and cyclones. Thunderstorms come mostly in summer, cyclonic storms may come at any time of year. Thunderstorms cover small areas while cyclonic storms may cover up to 4000 square kilometers. Thunderstorms can come from any direction while cyclones usually come from the west. Thunderstorms only last a few minutes or a few hours while cyclones may last for days. Also thunderstorms produce fast pelting rain while cyclonic storms often have a gentle slow falling rain.

The most severe storm is the tornado. These are characterized by funnel shaped clouds whose lower areas extend down and touch the ground. These storm tips are seldom over a kilometer in diameter but they cause many more deaths than any other type of storm. Their winds reach velocities of 200 to 300 knots. No wind instrument has ever survived the tornado but estimates of wind speed are made by damage and unusual phenomena such as sticks being driven through electric poles with the pole still standing upright. Tornadoes produce damage to houses by lowering the pressure around the house causing them to break apart from the pressure inside.

Spring and summer and especially May and June are the months with the most tornadoes in the United States. There are about 200 sighted tornadoes each year. They move from the Great Plains to the Mississippi Valley and north to the Ohio Valley. This is called Tornado Alley due to their frequency in this area. Tornadoes are not confined to this alley however and they have occurred in every state east of the Mississippi River. Tornadoes often occur ahead of cold fronts.

Hurricanes are tropical storms also known as typhoons. They have low pressure areas at their center, these are known as eyes. Hurricane winds are counterclockwise in the northern hemisphere and these reach velocities of up to 70 knots. Most damage during a hurricane is caused by the heavy rains from the warm moist air which promotes flooding.

Weather Maps

The United States maintains about 400 weather stations around the country. These are assisted by over 6000 volunteer stations. These stations turn in all types of weather data which are plotted on maps called synoptic weather maps. These are published daily and sold in many varieties to interested people at small cost. The Weather Bureau also has machines which print out maps at all hours of the day. An individual or a company can buy a duplicate machine and by setting it properly can receive an instant printed recording of the latest weather map.

The data on weather maps include temperature, dew points, pressure, cloud coverage, wind direction, wind speed, and barometric tendencies for each official weather station. The maps also include location of air mass centers, cold fronts, warm fronts, stationary fronts, occluded fronts, high pressure areas, low pressure areas, isobars, and areas of precipitation. Accompanying maps give isotherms, upper air flow, and possible weather predictions.

8

CLIMATES

Climate is the average of weather conditions over a long period of time. Weather data are amassed at a given location for several years and from these an average is compiled, thus the extremes of weather which occur infrequently are averaged out for any particular spot. Climates are influenced by latitude, land and water distribution, altitude, land barriers, air movement, ocean currents, permanent storm areas, and high and low pressure centers.

Latitude is a major factor in determining the amount of insolation a place receives during the year. The insolation of each place varies since the sun moves to different overhead positions constantly. When the sun is overhead at the Tropic of Cancer, the northern hemisphere receives the most insolation since the more nearly vertical rays shine on the north and there are more hours of sunshine. The reverse is true when the sun is over the southern tropic.

The distribution of land and water plays an important part in climates since land heats up and cools off faster than water. Because of this, inland climates are more extreme than sea coast climates. Much of the insolation value is lost to the sea since heat is used to evaporate water. This is later liberated in the atmosphere when condensation and precipitation occurs. When the sun's rays penetrate the ocean the water is heated or affected to depths of three to ten meters whereas the same sun on land affects the earth to about five to ten centimeters.

Altitude or height above sea level is also a major controller of climate. It affects both temperature and precipitation which are the basic elements of climate. A person traveling up the side of a mountain goes through the same climatic zones as if he were traveling toward the poles. Mountain areas are cooler than plateaus and plains areas since they radiate heat in all directions whereas a plateau or a plain only radiates heat upwards. Also mountains generally have cloud cover which prevents the warm sun's rays from heating its surface.

Land barriers help to contain and modify climatic conditions. Marine influences are cut off from the interiors by highlands which result in modification of temperature and precipitation. Areas which are on the lee side of mountains receive less precipitation as the result of moisture in the air condensing as it is lifted to higher elevations. Oregon and Washington which receives about 190 centimeters of precipitation on the coastal sides of the Cascade Mountains have desert conditions on their inland lee sides. Sunny Italy and Greece are that way since the Alps prevent the cold snow laden winds from the north from moving into the Mediterranean Sea area.

Air movements are one of the more noticeable regulators of climate. Winds moving from the equator bring warm air and winds moving from the poles bring cold air. Winds moving off water bring moisture laden air, especially if they are warm.

Closely allied to the air movements are ocean currents which can greatly modify climates. Cold ocean currents cause coastal climates to be cool and the air to be dry. Warm ocean currents cause coastal climates to be moist and warmer than they ordinarily would be. The Gulf Stream becomes the North Atlantic Drift as it leaves the Caribbean and heads out to sea toward the British Isles. It gives that area a more moderate climate than it would have at that latitude. Ocean currents are like rivers in the ocean, some colder and some warmer but easily distinguished by their differences of temperature than the surrounding water. Ocean currents usually follow the prevailing wind directions and therefore both work together to affect the climate of an area.

Storms occur in set patterns and are permanent fixtures in many areas. The coast of India and Burma can expect the monsoonal rains each year. The Cape of Good Hope is an area of storms. The Caribbean is an area of permanent summer hurricanes. Each of these play an important part in the climate of the land that they pass over.

Permanent high and low pressure areas are responsible for air moving out of an area or into an area. Permanent highs exist over Canada and Siberia, permanent lows exist over the equator and the Caribbean. Highs generally indicate unchanging weather and lows indicate changing usually stormy weather.

Classifying Climates

The word climate is derived from the Greek work klima which means to slant. The early Greeks realized that the slanting of the sun's rays was the dominant factor in climate control. From the time of the early Greeks, systems have been devised to classify climates thereby making them easier to understand. Having a classification system saves much time in communication when climates are discussed. For instance, the term "desert" conjures up images which would take many words to describe. This is a strong image whereas the term midlatitude monsoon probably has little meaning to most people. Thus it is necessary to classify the climates so that the less known types are more easily understood.

The German meteorologist Koeppen was the first to devise a universal classification system of climates which was acceptable to most scientists. His system survives today with little modification; however, slight changes in his system are needed for more accurate identification and discussion. The Koeppen system takes into account latitude, temperature, precipitation, and vegetation. Other factors are considered but are of minor importance.

The letters A, B, C, D, and E were first applied to latitude starting with A at the equator and proceeding through to E at the poles. Today the letters have come to mean more than a latitude designation. The letter A represents tropical climates, B dry climates, C midlatitude climates, D snow forest climates, and E polar climates.

The letters a, b, c, and d are temperature indicators. Those with hot summers are labeled a, those with warm summers b, those with cool summers c, and those with extremely cold winters as d. Thus a climate labeled Db would be a snow forest climate with a warm summer.

The letters f, w, and s indicate the precipitation patterns of an area. The letter f represents precipitation in every month, w indicates a dry winter, and s a dry summer. Sometimes the small letter m is used to indicate monsoonal conditions A climate designated Cfa would be a midlatitude climate with precipitation all year and with a hot summer.

Other letters may be attached for clarification. These will

DEPARTMENT OF NATURAL SCIENCES (SP)

be mentioned where they apply. For instance the letter H designates a highland climate. Since it would be almost impossible to indicate the variations of the highlands it is more convenient to label these H and it is assumed that it gets colder as altitude is increased.

Tropical Climates

There are three main types of tropical climates, these are (Af) tropical rainforest, (Aw) tropical savanna, and (Am) tropical monsoon. They have high temperatures and abundant precipitation in common although the precipitation varies during the year.

The tropical rainforest climate exists in the Amazon area of South America, the Zaire area of Africa, and the East Indies. It follows the equator. These rainforests are areas where the average monthly temperatures hover around 27° C and the precipitation is over 100 centimeters a year. The rainfall is well distributed and precipitation occurs almost every day. In this area large trees reach heights of over 70 meters. The floor of the rainforest is clear and travel on foot is easy. The forest is a gray green color and trees may be found in all stages of fruiting and blooming. Trees such as the rubber tree, banana, mahogany, teak, and various nut trees are found here. The rainforest vegetation is also referred to as selva.

Animal life in the rainforest consists mainly of large snakes and monkeys. There are other animals but they are in the minority. Meat is difficult to obtain here and fishing in the large rivers is a prime occupation. These areas have been historic locations of cannibalism probably due to the shortage of available meat for domestic consumption. Agriculture in these areas consists of cutting down the trees and burning the limbs off them. Crops are planted between the fallen trunks. The soil becomes exhausted very quickly due to the high rainfall and temperatures and in a few years the farmers must move on. This subsistence migratory agriculture is known as slash-and-burn farming. In Africa it is known as fang and in South America as milpa and in Southeast Asia as ladang. The main crop planted is usually cassava or manioc which is a small tree producing large starchy edible roots. The mechanically cut rainforest seldom recovers and there then follows great destruction to the ecology. This cutting is probably the greatest threat to the future of the rainforest.

The tropical savanna climate produces a grassland vegetation. It exists on both sides of the tropical rainforest and blends into it by producing small trees which get to be larger trees as the rainfall increases. These grasslands exist in Brazil and Venezuela in South America, from Panama to Southern Mexico in Central America, in central Africa on both sides of the rainforest, in central India, in most of Indochina, at the northern fringe of Australia and the tip of Florida. The Everglades are the American savanna.

The savanna receives adequate precipitation in the summer but in the winter dry season, life is miserable for those living under near primitive conditions. Brazil experiences its drought in July and August while Nigeria has its dry season in January and December which is their respective winter months.

The savanna is the home of the big game animals of Africa. Here roam the lion, giraffe, zebra, rhino, and the elephant. In other savanna climates the animals are more deerlike and all are sought for food.

The tropical monsoon climates exist in the Guiana Highlands of South America, the west central coast of Africa, and on the west coasts of India and Burma. These would be classified as rainforest climates except that in two summer months of the year there is an inordinate amount of precipitation. Cherripunji in India has received over 2500 centimeters of precipitation in a year.

The tropical rainforest and savanna areas of South America and Africa do not contain large populations. The problems in these areas center around controlling growth of vegetation in the rainforest and supplying water during the dry season in the savanna. Infectious diseases are numerous due to the high humidity and temperatures. Also in these climates, insects make life unpleasant for sedentary humans.

Dry Climates

The B climates are of two major varieties, BW which is desert and BS which is steppe. The desert climate receives less than twenty usable centimeters of precipitation in a year and the steppes receive between twenty and forty.

Deserts are tropical or topographic. The tropical deserts exist where descending air dries the land and prevents adequate precipitation. These are the Tropic of Cancer deserts of Sonora in North America and the Sahara and Arabian in Africa. The southern tropical deserts are the Atacama of South America, Kalahari of Africa and the Victorian of Australia.

Topographic deserts are the result of mountain barriers and wind circulation. The largest of these is the Gobi Desert of Mongolia. Topographic deserts usually have temperatures well below freezing in their winters. All deserts have great diurnal extremes of temperatures.

Deserts are characterized by xerophytic vegetation, that is, vegetation which have systems that conserve moisture. Among these are the creosote bush and cacti. Animal life consists of turtles, rattlesnakes, and nocturnal mammals and birds. Some deer live in desert areas but these are not large in numbers. Humans living in desert areas are usually engaged in raising sheep and goats for their sustenance.

The steppes are grasslands. These are the wheat and cattle raising regions of the world. The steppes receive enough precipitation for grass but not quite enough to support trees. The steppe climate exists on the Great Plains of North America, the Ukraine of the Soviet Union, the Pampas of Argentina and on the fringes of all desert regions around the world. In years when the precipitation nears forty centimeters the wheat and cattle increase in production and in bad years, that is when the precipitation nears twenty centimeters there is disaster to the agricultural economies. Dust storms blow the soil away from plowed fields in the dry steppes. Cattle hooves increase the damage by destroying the grass cover. This too is easily eroded by the winds. Irrigation has made many steppe areas of the world into prime farm land but there is danger from salts which accumulate on the surface due to the evaporation of irrigation waters.

When warm moist tropical air is occasionally brought onto the continental interiors strong convectional storms arise causing cloudbursts in these dry areas. Erosion is rapid and much vegetation is destroyed by the resulting flash flooding.

Midlatitude Climates

The C climates are middle latitude climates, they are varied in their vegetation forms and in their utility to man. The bulk of the world's population lives in C climates. The major climates of these types are (Cfa) humid subtropical, (Cfb) marine west coast, (Cs) mediterranean, and (Cw) midlatitude monsoon.

The C climate areas are the areas of cyclonic wind movement. They are alternately affected by polar and tropical air masses. The particular type of climate depends upon the dominant air mass of the region. Humid subtropical regions are influenced mostly by maritime tropical air masses, marine west coast by marine polar air, the mediterranean by both continental and marine tropical air, and the midlatitude monsoon by continental polar air.

As indicated by its symbolic designation the humid subtropical climate has precipitation in every month of the year as well as hot summers. These are areas of pine and deciduous forest and when these are cleared they become areas of good producing farmland. The humid subtropical climates exist on the southeast coast of continents. In North America it exists from Florida north to New Jersey and west to the steppe areas of the Great Plains. Central China and northern Argentina and Uruguay have this climate as well as eastern Australia.

The marine west coast climate is cool and usually moist. It exists in the Pacific Northwest of the United States and covers most of Europe and all of the British Isles. These areas have moderate summers. The Pacific Northwest is the land of the giant Douglas Fir tree and is America's biggest lumber producing region. Generally forest would cover these areas around the world but humans have lived in them so long that most of the area is now under cultivation.

Mediterranean climates exist in the Mediterranean Sea area, around the state of California, in Chile, South Africa, and the southwestern tip of Australia. These are areas with dry summers and moderate precipitation the rest of the year. Vegetation is in the form of bush type trees, among them olive, citrus fruits, figs, pomegranates, cork oak, and palms. Grapes are a premium crop. The famous grape wines of the world come from the mediterranean climates of Italy, France, Spain, and Chile.

The midlatitude monsoon is a winter monsoon. It is created when high pressure areas form over Siberia in winter and move toward the Pacific Ocean. This cold drying wind moves over northern China and Japan giving them an extremely dry winter. Areas affected by the dry monsoon are usually grasslands since trees cannot withstand the drying winds.

Snow Forest Climates

These are so named since the snow which falls in winter lasts on the ground for about a hundred days and the vegetation is forest. These D climates are of two major varieties (Da,Db) humid continental and (Dc,Dd) subarctic. The D climates do not exist south of the equator since there are no large land masses to create the extreme variations necessary for snow forest conditions. The highest extremes of temperature are found in the snow forest climates, these vary from 40° below zero Celsius in winter to near 50° C in summer.

The humid continental areas stretch from northern United States into southern Canada and from Norway across Europe into southern Siberia of Asia. The humid continental is sometimes divided into cool and cold varieties depending upon whether they have hot or warm summers. These climates produce trees and mark the northern limit in which fruit trees such as apples and pears can be grown. The humid continental climates have large populations in their southern areas, such large metropolitan areas as Moscow, Cleveland, and Toronto are in this climate zone.

The subarctic climates are found below the Arctic Circle and are areas covered with trees. Such trees as spruce, maple, aspen, birch, larch, and tamarack are found here. These are the winter headquarters of the reindeer. Moose, elk, and bear also live in the D climate forests, especially in the subarctic variety.

The snow forest climates have sparse populations when the vast areas covered by them are considered. These are the major lumber regions of the world. Norway, Sweden, Finland, the Soviet Union, and Canada consistently have large lumber production and large exports of lumber products. Some forest areas of the subarctic are referred to as Tiaga.

Polar Climates

The E climates exist near the poles; they are of two basic types (ET) tundra and (EF) frost. The frost climates have permanent ice and snow cover. These are found in Greenland and Antarctica. Humans occupy these areas mostly out of scientific curiosity rather than with an eye to permanent habitation.

The tundra is a grassland extending over the northern fringes of North America and Asia. Tundra also exists on the coasts of Greenland and as a small fringe on Antarctica. The tundra has not only grasses but lichens, mosses, lings, and sedges growing on its thin soils. Much of the tundra is barren gravel due to strong winds blowing away the thin soils.

The tundra is the home of the Lapp and the Eskimo. It is also the summer home of the musk ox, moose, elk, and reindeer. The tundra is in the land of the midnight sun and the winters here are monotonously long. These are outpost stations in a civilized world and attempts at agriculture, even in greenhouses, usually have ended in frustration.

Climates and An Expanding Population

As humans look about for areas to which the expanding population can be moved they cannot help but notice the vast areas covered by the tropical climates, the deserts, the subarctic, and the tundra. The problem in the tropics is controlling growth and making the temperatures and moisture bearable. Experiments with farming and plantation management have not been too promising.

In the deserts, the problem is one of adequate moisture for domestic and agricultural consumption. Since sea water cannot be used directly for agriculture due to its dissolved salts, fresh water must be obtained in large quantities. Where streams flow through desert regions they are probably utilized to their maximum extent and some other source of water must be made available. Distillation of sea water is possible but the present cost is prohibitive.

The other areas of sparse population, the cold lands, could probably support more people if adequate housing were available and food could be brought into the area inexpensively. These areas will probably not be self sustaining unless drastic new agricultural and fuel discoveries are made. Entire new concepts of living will have to be developed in order to make life pleasant in the subarctic and tundra for midlatitude peoples.

9

EARTH FORCES AND ENERGY

Forces which mold the earth may be divided into those which operate within the earth and those which operate upon the surface of the earth. The internal forces which mold the earth are termed endogenic. Of these volcanism and diastrophism are dominant. These cause rocks to be bent, broken, and expulsed. The surface or exogenic forces operate under the influence of the sun and gravity. These forces find their expression in the activities of wind, water, ice, and living organisms on the earth's surface.

Energy

All earth processes and forces are the result of energy transformations. Energy is the capacity for producing or retarding motion. The Law of Conservation of Energy states that energy cannot be destroyed but only changed in form. This does not mean that energy cannot be lost to the earth for it often is.

Energy is tied up in the earth structure in many forms. It also comes to the earth from the sun. Earth processes operate as the result of transfers of energy upon and within the earth.

Potential energy is stored energy, it is the energy of position, it is a manifestation of gravitational attraction. The light fixture over your head has potential energy. It has the capacity of movement and doing work. If the light fixture weighs twenty kilograms and is five meters overhead and it should suddenly break loose it will move twenty kilograms a distance of five meters, thus doing 100 kilogram meters of work.

Kinetic energy is the energy of movement. Every moving object possesses and releases kinetic energy. It is the velocity of the moving object times the mass of that object. As the light fixture begins to fall it has a maximum of potential energy and a minimum of kinetic energy. Just before it hits the floor it has a minimum of potential energy and a maximum of kinetic energy. Where does the energy go after it hits the floor?

The earth is constantly storing solar energy or energy from

the sun. This energy evaporates water and moves the air. It is stored in plants by the process of photosynthesis and in the animals that eat the plants. This energy eventually returns to the land in the transformations of chemical energy. Or the solar energy may be tied up as fossil fuels deep within the earth to be released when utilized for human endeavors.

Heat energy is a manifestation of the kinetic energy of molecules and atoms. When heat energy is applied to matter the matter's molecules vibrate faster and they slow down as this energy is expended.

Chemical energy is the energy of the formation of compounds. It is the force which holds atoms in union. It is the energy gained or lost when atoms have lost or gained electrons. Electrical energy is created when electrons move from one part of a substance to another. If it moves in a set direction it is referred to as a current. Atomic energy is the binding energy which holds the nucleus of the atom together. When the atom is broken apart tremendous amounts of energy are released in many different forms, including heat energy, light energy, and movement of air which is mechanical energy.

All matter and the changes of matter may be considered as energy and energy transformations. The earth and everything on it is in a constant movement and change. All change is the result of energy transformations.

The Effect of Rotation

Various forces and effects are set up by the earth's rotational movement. Differences in speed of rotation exist due to location and distance from the axis of rotation. Particles located at the poles have no rotational velocity while those at the equator rotate at a speed of approximately 1,660 kilometers per hour. Particles or people located at 30° North Latitude rotate at a speed of about 1,440 kilometers per hour. Those at 50° North at a speed of 1,070 kilometers per hour. In other words, there is a maximum rotational speed at the equator which decreases as one approaches the poles.

Rotational velocities set in motion the centripital and centrifugal forces on all parts of the earth except at the poles. The centrifugal force tends to throw a rotating object off into a straight line. It is similar to a weight swung on a string. When the string is released the weight travels in the direction of release. The force which previously acted on the string holding the weight in a circular path is centripital force. This force is directed at the axis of rotation. Therefore the centripital force is reduced as one nears the poles since it is closer to the axis of rotation so the two directions of force must not be confused. Also since the centrifugal force, the force which tends to pull away, is only an apparent force it is sometimes thought of as a negative centripital force.

In a centrifuge, the centrifugal force separates light materials from heavy materials by throwing the lighter particles to the outside. As earth materials are whirled the lighter materials, the gases, move to the outside and heavier metallic materials move to the inside. With rock materials the lighter rocks make up the continents and mountains while heavier rocks make up the ocean basins and mountain roots.

An apparent effect of rotation on moving bodies on the earth is the Coriolis Effect mentioned earlier. Basically it is this, moving objects such as wind in the northern hemisphere are deflected to the right while moving objects in the southern hemisphere are deflected to the left. To understand the full implications of the Coriolis Effect is difficult without intricate mathematics. The effect is due to the differences in angular velocity of places at different latitudes on earth. A particle traveling from the north pole along a line of longitude slows down in angular velocity since its distance from the axis of rotation is increased and therefore falls behind or to the west of its original longitudinal line. A particle traveling northward in the northern hemisphere from midlatitude toward the north pole along a line of longitude must increase its angular velocity to be in equilibrium because its distance from the axis of rotation is decreased therefore it moves to the right or east of its original longitude. This increase in speed as the distance to the axis of rotation is decreased may be demonstrated by tying a small weight to a string and swinging it in a circle. Let the string wind around your finger without moving your finger. You will notice that its angular velocity increases as the string gets shorter and the weight gets closer to the axis of rotation.

More difficult to explain is the deflection to the right of bodies moving along a line of latitude from east to west or vice versa.

It is understood when one considers that the moving body is moving in a straight line but the surface from which it is being measured is curved.

As the velocity of a moving body increases so does the Coriolis Effect, that is, if the speed of a moving object is doubled so is its deflection. To repeat, as an object moves toward the equator it moves further from the axis of rotation and therefore the Coriolis Effect is reduced.

Gravity

Gravity is the attraction of one body to another. Newton had stated that the force of attraction between two bodies depends upon their mass and the distance between them. Gravity of the earth is expressed in weight, it is a relative measurement and we use arbitrary comparisons such as a kilogram to express it.

Gravity is best understood by the concept of acceleration of gravity which is the increasing speed at which bodies fall toward the center of the earth. Light bodies fall at the same speed as heavy bodies. The acceleration of gravity for bodies falling in air is 980 centimeters per second per second. This means that a body starts out falling 980 centimeters for the first second 1960 centimeters for the second second, 2940 centimeters the third second and so on.

Since the earth is not a perfect sphere, gravity differences exist. One is closer to the earth's center at the poles due to rotational flattening and would therefore weigh slightly more there than at the equator. Where the earth is rough such as in mountainous areas gravitational differences also would exist in small zones.

Magnetism

Almost everyone has worked with compasses whose function is based upon the magnetic forces within the earth. When lines of force have been set up in a suitable needle and it is allowed to swing freely it will align itself in an approximate north-south direction. This also occurs in nature in a form of magnetite called lodestone, which is an iron oxide which has become magnetized. Locations of ancient north magnetic poles have been determined by the orientation of magnetite grains in hardened lava. Liquid lava would have no orientation and no value for this purpose, but as the magnetite hardens it floats freely in the liquid lava and is able to orient itself with the magnetic poles. It is imprisoned in this orientation when the lava completely hardens.

There is a north magnetic pole located in north central Canada just north of the Arctic Circle and a south magnetic pole located between Tasmania and Antarctica just south of the Antarctic Circle. These do not line up as a magnetic axis, they are not opposite each other as poles. Neither are they lined up with either of the earth's rotational poles. Since the earth poles have migrated during the past they occasionally have overlapped the magnetic poles.

If the north magnetic pole is designated as positive and the south as negative, polarized magnetic objects on earth will align themselves with their negative pole to the north since opposites attract. It is best to think of the needle as having a north seeking pole since that portion has been often designated as the positive pole.

The magnetic force is strongest at the magnetic poles and weakest at the magnetic equator. Magnetic force differs for different locations of the earth due to differences in roughness of the surface and densities of earth materials.

Magnetic declination is the differences measured between the magnetic pole and the geographic pole. For most purposes magnetic declination is given in angular degrees of differences between magnetic north and geographic north. The magnetic declination varies from year to year due to the shifting of the magnetic poles. This forces surveyors to correct their instruments for magnetic declination each time they prepare to measure and record field data.

The U.S. Navy is one of the many organizations which is constantly measuring the earth's magnetic field. They measure the shifting depth of the pole within the earth, the intensity of the field, the horizontal and vertical intensity as well as the north-south and east-west intensities.

It is believed that the earth's core is liquid or metal in a plastic state and it moves slowly thus setting up currents which create the magnetic poles. This theory is further supported by convectional heat measurements within the earth. The fact that variations in convectional movements occur from time to time causes variations in the location of the earth's magnetic poles. This could also explain the phenomena of magnetic storms which are bursts of fluctuations in the intensity of the magnetic fields.

10

EXTERNAL EARTH PROCESSES

The earth's exogenic or surface forces operate under the influence of gravity and the sun. They are the forces of gradation which tear down high spots of the earth and fill in the low spots. As the internal forces tend to raise portions of the earth higher the external forces tend to lower them. Even though the earth is in a constant fluctuation there is still an inner balance. This inner balance of the earth is referred to as isostacy.

Degradation and aggradation take place in four steps. The material is first weathered and prepared for removal, then it is removed by some agent which is the process of erosion. Next it is transported by that agent to another location, and finally it is set down or deposited by that agent. Common agents of erosion and deposition are ice, wind, and water. Removal and deposition can take place merely by the influence of gravity and without the aid of an agent, one such movement is landslides.

Landslides Or Slope Failure

Landslides are the result of materials which were once in equilibrium responding to new conditions. They are best studied according to their speed of movement. They can be divided into those which move rapidly and those which move slowly. Of the fast movements some are enhanced by rainfall and others by earthquakes or weaknesses in structure. The erroneous view is often held that rainfall lubricates landslide materials. Those landslides enhanced by rainfall occur because the materials are no longer in equilibrium since water is a friction fluid and does not lubricate the hillside. The landslide occurs because the water of a rainfall adds more weight to the hillside. Some specific types of rapid landslides which are the result of the addition of water are earthflow, mudflow, and debris avalanches.

An earthflow is a type of landslide where large amounts of surface material slide down the slope causing hummocky topography. These occur in humid regions of hill country as well

DEPARTMENT OF NATURAL SCIENCES (SP)

as around steep road cuts. Another type, mudflow, is a soupy mass of mud which has collected in valleys of arid regions. The mud oozes slowly but perceptibly down slope. These can be quite large and involve many thousands of cubic meters of material. The debris avalanche is a water saturated landslide which consists of masses of surface materials which tumble over steep slopes.

Fast movements which are the result of slope failure with water playing only a minor role include rock slides, debris slides, and slump. Rock slides move thousands of cubic yards of materials in high mountain areas. This mass usually moves as a unit. These movements are due to many causes, among them, weak rocks alternating with strong rocks, steep slopes, mining activities, and faulting. A rock slide along the Gros Ventre River of Montana in 1925 moved 40 thousand cubic meters of material down into the valley and halfway up the other side. Most destructive landslides are of this nature.

A debris slide is similar to a rock slide but does not move as a unit. It mixes the surface and rock materials as it moves. Debris slides are caused by slope imbalance often associated with fissures in bedrock.

Slump is a movement of material as a unit along curved planes of weakness in soil or rock. Slumps cover relatively small areas and are found around coastal cliffs and cliffs of mesas. As the material moves down slope it twists in a backward rotation. Thus a tree which grew vertically will be found at an angle on a newly slumped surface.

Slow movements are creep, talus, stone polygons, solifluction, and rock streams. Creep is the imperceptible movement of materials down slope. The materials are small and the movement not dramatic; however, everything is on its way downhill and creep probably accounts for more tonnage moved downslope than the dramatic landslides. Evidences for creep may be found in electric poles and grave headstones pointing downslope after a few years in place.

Talus is the material which collects at the base of steep cuts and cliffs. After breaking off, heavier materials are at the top of the talus slope since the finer materials are carried to the base by rain. Talus cones are found around the small valleys which exists high on mountainsides. These are aided by small amounts of water which periodically flow in the valleys.

Stone polygons are the patterns of stones created by alternate freezing and thawing of ground in cold regions. Since the ground thaws slower under a rock than when exposed to air the rocks are raised above the thawed surface. The rocks then fall from this raised position into the polygon patterns which also gives them the names of stone nets and stone garlands.

Solifluction also operates in cold regions, mostly in regions where the subsurface is permanently frozen. In these permafrost areas the upper surface becomes a soupy mass sliding on the frozen underlayers. These ooze down into the lower areas and thus gradation occurs.

Rock streams are masses of rocks in valleys which are slowly working their way down slope. They are also called rock glaciers since their outer edges assume the lobate shape of a glacier front. These exist in valleys from Mexico to Alaska. No satisfactory explanation has been given for their collection and movement. They may be the dying stages of glaciers, or moved by water, or the final stages of debris avalanches.

Weathering

Weathering is a response of materials once in equilibrium to new conditions, usually new conditions of the atmosphere due to new exposure. It is a process by which large materials are made smaller, heavy materials are made lighter, and unstable materials are stabilized. The process of weathering makes erosion and removal of materials easier. There are two types of weathering, physical and chemical. Physical weathering involves making bulky materials smaller without changing their chemical composition, therefore physical weathering is disintegration.

Physical weathering takes place in rock materials which were once underground and covered with deep overburden. As erosion takes away the overburden the pressure is lowered upon the solid rock and the rock splits and cracks as it reacts to the release of pressure. Physical weathering also takes place when cracks in rock materials allow water to seep in and freeze. This formation of ice causes expansion and frost heaving which cracks the rocks further.

Physical weathering can also occur due to hot and cold changes. Since rocks are composed of different elements which have different rates of expansion and contraction they are affected by cold and heat and these tend to flake off rocks in layers or in grains. This process is exfoliation when slabs break off and in the case of flaking grains it is granular exfoliation.

Whereas physical weathering merely makes particles smaller chemical weathering actually changes the composition of the materials affected. It is decomposition since it involves breakdown. Chemical weathering is enhanced by heat, moisture, and chemical gases of the atmosphere.

Examples of chemical weathering can be seen by breaking open a rock and comparing the differences in color of the outer portion to the interior. Rusting iron is an example of chemical weathering. When limestones are washed away it is usually due to the formation of the bicarbonate ion which is a change from the normal carbonate ions of limestone.

Chemical weathering dominates in warm humid climates while physical weathering dominates in cooler areas. There would be a maximum of physical weathering at the poles and a maximum of chemical weathering in the tropics.

Soils And Their Formation

Soils are a direct reflection of the processes which have gone to create them. The study of soils, pedology, is the study of weathering since weathering is the major process in soil formation. Soils are influenced by temperature, precipitation, native vegetation, drainage, lithology, and organic activity. To understand soils and their formation is to understand the majority of earth processes.

On a world scale the important soil concept is that of zonal soils. These have good development of a soil profile from parent material. A soil profile is a cut through the soil from the surface down to the parent material or the material of the soil's origin. The profile is divided into four horizons, these are labeled A, B, C, and D with A being at the top and D at the bottom. The A horizon consists of humus materials from which minerals are leached and carried away or down into the lower layers of the soil. The A horizon is the topsoil. The B horizon is usually the recipient of the leached materials and the zone of accumulation for these materials. It consists of clay and slightly weathered materials. The B horizon is the subsoil and is usually lighter in color than the A or humus layer. The D horizon is the parent material which may be stream gravel, glacial deposition, sand, rock, or any other material from which soil may be formed. The C horizon is the transition from the D to the B, it is a layer of disintegration and decomposition. If the parent material is rock then the C horizon consists of large rocks at its base and smaller more weathered rocks near its top, these grade into the particle sizes found in the B horizon.

Zonal soils form by three basic processes, podzolization, calcification, and laterization. The factors which govern the processes are climate, topography, the parent material, the vegetation of the area, and the length of time without disturbance.

Podzolization occurs in moist cool regions such as in eastern United States and Canada. The soils created are called pedalfers due to their high content of aluminum and iron. The soils are acid and are influenced by the decay of leaves and needles of trees. These are forest soils and have various identifying names. Pedalfers stretch from New England south to Florida and from the Atlantic west to the Great Plains. The differences in the divisions of pedalfers are due to differences in temperatures of formation.

Calcification creates pedocals which are high alkali soils due to their concentration of calcium carbonate or lime. These soils are built up from grasslands under arid and semiarid conditions. They predominate in western United States. These include the Chernozems of the Great Plains and the Sierozems of the deserts. The differences in pedocals are mostly due to differences of precipitation.

Laterization creates laterites which are tropical soils. These are created by high temperatures and high rates of precipitation. In these soils bauxite an aluminum ore and hematite an iron ore are concentrated. These are soils of the West Indies, Central America, Brazil, and Venezuela.

If soils have one controlling factor in their environment and it dominates their formation they are intrazonal soils. Such soils are bog soils and the soils of salt flats. If the soil has no zonation it is azonal and these involve among others dry sand and alluvial fans. If they become wet then zonation begins.

Underground Water

About a sixth of the precipitation in humid areas seeps underground. This is subterranean water or subsurface water. It exists in the pore spaces of rocks as well as in joints and crevices. The water is usually near the surface. The upper surface of this water is the water table and it takes the shape of the land. As the water moves from one place to another it dissolves rock materials and deposits them elsewhere. It is this solution of materials which results in the formation of the cave or spelean environment.

Regions which are affected by underground water solution usually reflect it in surface features. These regions are called karst after a region in Yugoslavia similarly affected.

The surface depressions which lead down to the caves are sinkholes. They take various forms due to the amount of water and the density of the rock in the area. If the sinkhole is funnel shaped it is a doline, if rectangular it is a ponor, and if the cave collapses leaving a valley on the surface it is a uvala. Uvalas usually contain disappearing streams.

Once underground solution nears completion the holes enlarge to the point where there are remnant hills which are then the dominant landscape feature. These are known in different parts of the world as haystack hills, hums, magotes, and pepino hills. They are erosional features just as the sinkholes are erosional features.

In order for karst regions to develop, at least moderate rainfall must be present, also a soluble dense rock such as limestone must exist and the rock must be highly jointed and preferably slanted. The water which enters the rock and dissolves it must be drained off to a lower area so that there is a continuous removal of the dissolved rock materials.

Deposition in karst regions are in the form of cave deposits. These are stalactites which hang like icicles from the ceiling and stalagmites which protrude upwards from the floor of the cave. When these join they form a column. Petrified wood is also an underground water deposit. The wood fibers are removed and replaced by precipitates, usually forms of silica and calcite. The siliceous mineral is most often opal and the calcitic mineral is most often travertine. Opal is also abundant around hot springs and geysers which are underground waters brought to the surface by steam pressures. The general terms flowstone and dripstone are used to describe and indicate cave deposits.

Running Water

Running water creates most of the landforms of continents. Except for the solid icecaps at the poles running water occupies all continents. It washes the slopes and takes the materials to the valleys where the main streams carry the materials to an area of first deposition and eventually the ocean. Continental landforms are the expression of the water which falls upon them. Running water in association with other processes or by itself governs the shape of the land.

The major stream of an area and its tributaries are a river system. The lowland occupied by the stream is a valley and the area between the streams is an interfluve. The entire area drained by a river system is its drainage basin and the area separating two drainage basins is a divide.

The speed at which a valley is deepened depends upon the velocity and volume of the stream. The velocity depends upon the gradient which is the slope of the stream bed. High gradient streams are swift and these can cut down rapidly since the speed of the stream determines the size of particle the stream will be capable of carrying. A stream traveling at a kilometer per hour will carry particles of coarse sand, at two kilometers per hour it can carry particles the size of pebbles and a stream moving four kilometers per hour will carry fist sized cobbles. If a large stream had a high velocity it would lower the landscape in a very short time period. However, most large streams have low velocities.

The lowest point to which a stream may erode its channel is its base level. The ocean is the ultimate base level since no stream can erode lower than the ocean. Temporary base levels may exist as lakes into which streams empty, but these will eventually be destroyed by the stream as it moves to the ocean. Another temporary base level may be created by resistant strata but streams eventually cut through these.

There are many ways of classifying streams. One of the most useful methods is to classify streams according to stage of development. Streams in this classification are in either a stage of youth, maturity, old age, or rejuvenation.

Youthful streams have V shaped valleys, steep waterfalls, rapids, gorgelike ravines, high gradients, undrained interfluves, and a small number of tributaries on each main segment. Mature areas have many main streams with a large number of tributaries. All land is in slope and there are no undrained areas. The valleys are broad with some small meandering and with the beginning of flood plains.

Old age landscapes have no falls or rapids. There are a few main streams with very few tributaries. They have strong meandering and large flood plains. The meanders of old age streams eventually develop ox bow lakes. If the stream continues without interruption into its final stages, a peneplain is developed. This is a large level plain near sea level. Some resistant hills called monadnocks may be left on the peneplain due to favorable location or due to their superior resistance. Since there are no identifiable recently formed peneplains available for study the theory of monadnocks may never be proved.

Most streams reach the stage of old age and then are uplifted before peneplanation. These interrupted streams have the meanders of old age but their valleys are steep sided as in youth. These meanders are said to be entrenched. Interruption in the cycle of erosion may be the result of glaciation, crustal movements, sea level changes, or changes in climate.

Streams erode their channels by plucking, by abrasion, by solution, and by lifting. Most stream erosion takes place in a very short period of time, usually during the flood stage or during the spring when waters are abnormally high.

The eroded material is carried by the stream in solution, in suspension, or it is rolled or skipped along on the bottom as bed load. In low seasons the streams will precipitate the materials carried in solution. These may be evident as powdered deposits on exposed rocks. Materials in suspension are deposited as the velocity of the stream changes. Bed load is moved only when sufficient volume and velocity are obtained.

The capacity of a stream is its ability to carry material and its competency is the largest particle size it is capable of moving. The stream's load is the actual amount of material being carried. Load is always less than capacity.

Streams constantly move to a lower level leaving behind terraces above the stream. Paired terraces are at the same level on each side of the stream. These may be depositional in nature or erosional, if erosional they are called straths. If a stream meanders, the terraces are unpaired. The stream moves toward the outside of the meander causing the slope above the outside to be steep and undercut. The stream slips off the inside of the other slope opposite the undercut cliff. Deposition occurs on the inside of the meander, at the bottom of the slip off slope.

The sand and gravel carried by the streams are deposited in bars and deltas. They also form floodplains when deposited on open areas. When streams overflow their banks they lose velocity and therefore their carrying power. Thus streams which overflow their banks deposit material on the edges of their banks forming natural levees. The nonstream side of the levee is usually swampy and is referred to as a backswamp. If the water of the swamp is capable of flowing then a stream will be created which will parallel the main stream keeping separated from it by the natural levee. Such streams are Yazoo Type streams getting the name from that river in Mississippi.

Materials carried by intermittant streams in dry areas create alluvial fans or alluvial plains since the arid area streams rarely reach the ocean. In the low areas the dry stream beds are termed arroyos, dry washes, wadis, or barrancas.

Marine Waters

Ocean waters have vertical and horizontal movements. The vertical movements may be pelagic or tectonic. Pelagic movements are caused by the addition or subtraction of water in the ocean, usually the result of expanding or declining glaciation. Tectonic movements are those indicated by the land rising or sinking in relation to the ocean. These may result from increases in elevation of the land, changes in elevation of parts of the ocean floor, or expansion or contraction of the distances between separated land masses. A wave on water has vertical movements but these are not considered since the main direction of energy of a wave is horizontal. The particles transmitting the wave energy merely rise and fall in place.

The horizontal ocean movements are currents. These have many origins. There are thermal currents, longshore currents, tidal currents, wind drift currents, and density currents.

The thermal currents are identified by their differences in temperature compared to the ocean in general. Current directions are associated with the Coriolis Force and the prevailing wind movement. Warm currents move toward the poles and cold currents move toward the equator. In some areas of the world, cold water is brought up from great depths by the shape of the ocean basin and the rotational direction. Such an area is found off the coast of Venezuela, this however is not a current but upwelling.

Wind drift currents are those created by the prevailing winds. They build up waves as they travel in toward the land. When the waves break they become turbulent surf. This surf creates shore features. As the broken waves return back to the ocean they may create undertow which is a stream flow away from shore. These undertows are dangerous to swimmers since many undertows are powerful. A swimmer caught in an undertow is advised to swim parallel to the shore instead of trying to fight the undertow by swimming back to the shore. By swimming parallel to the shore the swimmer swims through the undertow stream as if he were crossing a raging river.

When waves break on shore at an angle they create a current parallel to the shore and against the shore. These are longshore currents and they are responsible for moving most of the shore materials along the coast.

Tidal currents are caused by the attraction of the moon and the sun on the ocean waters causing them to bulge. When the moon is between the sun and the earth the tides are highest and when the earth is between, the tides are lowest. When the moon is at quadrature the tides are average tides known as neap tides. The term ebb tide refers to a retreating tide. Generally the height of tides is determined by the shape of the land upon which the water is spilling. Narrow inlets have higher tides than open shore. When the tide approaches a river mouth it comes up the river as a wall of water called a tidal bore.

Density currents are due to turbidity or differences in salinity. Turbidity currents are created when oceanic landslides place sediments in suspension. These currents have much force and move rapidly.

Density currents created by salinity differences exist where evaporation is high. In the Mediterranean Sea the salinity is higher than in the Atlantic Ocean since evaporation is higher in the dry mediterranean climate. Hence water from the Mediterranean flows into the Atlantic at low levels since it is heavy and the lighter Atlantic waters move in at the upper levels in order to bring the density differences into equilibrium.

Seismic waves are large sea waves created by earthquakes under the sea. These waves have improperly been called tidal waves since they have no connection with tides. They are properly termed tsunamis. When these batter a shore they cause much destruction to villages located there.

The waves generated by the various currents create the coastal shore features. Cliffs are created by undercutting, sea caves by cutting in, and sea arches by cutting through. Sea stacks are the remnants left behind when the arches collapse. As waves pound and level the shore they create wave cut terraces.

Depositional features by marine action are in the form of beaches and bars. Beaches may be sand, stone, or cobble. Stone beaches are more common than sand beaches but bathing beauties do not lie around on stone beaches and therefore these beaches are seldom photographed. Curved sand bars on the edges of protrusions of land are referred to as spits or hooks. If they occur offshore they are barrier bars or if in front of a bay they are bay barriers. A sand bridge between two solid land areas is a tombolo.

Deposition in oceans can also occur as a result of organisms. The best known of these is coral which creates reefs. Coral deposited around a land mass is a fringing reef, those offshore are barrier reefs, and those surrounding a central lagoon are atolls. Deep ocean organic deposition is fine and mostly microscopic. There are large deposits of organic matter called oozes. The largest of the oozes in area covered is that of globigerina, a single celled protozoan, second in abundance is the radiolarian ooze also created by a protozoan. Next is the diatom ooze created by a plant and fourth is the ooze of the pterapod a minute crustacean.

The major sediment of the deep ocean is red clay which far exceeds the oozes in area covered. The origin of the clay is not known but theories suggest that it may be dust from the dust bowls of the world, or it may be volcanic dust, or it may be dust from outer space, or it may be precipitated by some unknown process.

There are also isolated deposits of nodules and sands in the deep ocean, these seem to originate by precipitation. Manganese nodules the size of large eggs are found at depths of seven hundred meters in the deep ocean. There are also sand sized particles of glauconite of some significance in several pockets of the ocean floor.

Glaciers

Glaciers are large masses of ice which have been created from snow and which move under the influence of gravity. Snow falls upon high mountains and in high latitudes and may remain there permanently. In fact, snow falls over most of the earth. Most of the snow melts but in the extreme poleward regions and at high elevations the build up of snow often is faster than melting. As the snow builds up it forms pellets called neve or firn. These are in reality sedimentary particles. When compressed they form rock ice. When the ice has reached sufficient thickness it may become unbalanced and start to move, once it starts to move it is a glacier.

Those glaciers which cover large land areas such as Antarctica and Greenland are ice sheets or continental glaciers. Those that originate in the valleys of mountains such as in the Alps or in Alaska are alpine or mountain glaciers. When the mountain glaciers move down to the lowlands and meet they form piedmont glaciers.

Mountain glaciers originate in amphitheater like basins high on the mountain slopes, these basins are cirques. They are created and made larger by the ice movement out of this location. The crevice between the glacier and the upper headwall of a cirque is a bergschrund. After the ice melts the cirque usually fills with water creating a lake called a tarn. As the glacier moves down the valley it creates smaller basins in chain like fashion. These, when filled with water, are paternoster lakes and are named for their resemblance to prayer beads.

Cirques will eventually eat back into the mountain chain causing a saw toothed ridge to be formed. These ridges are aretes and the passes between them are cols. If an isolated peak is attacked on all sides by glaciers it becomes a horn, the Matterhorn of the Alps is the most famous of these. As the glacier moves down the valley it changes the V shape to a U shape which is characteristic of valleys attacked by mountain glaciers. Since these are then lower than the original valley there are often many hanging valleys emptying into it. Hanging valleys are the remains of valleys which once joined the main valley at stream level. The areas between hanging valleys become truncated spurs with a flat face toward the valley. Where the glacier has moved over rock it may produce small ridges termed rouches moutonnee named after these features of the Rhone Valley of France. Large embossed rock features such as those found in Scotland are crag and tail.

Deposits by glaciers are called till when put down by ice and when deposited by the streams flowing from the ice are termed outwash. Tills are named according to their major component, thus there are sandy tills, cobble tills, clay tills, and boulder tills. If the till position in the glacier can be identified then they are moraines. There are terminal moraines deposited by the leading edge of the ice, lateral moraines at the sides, and ground moraines under the ice.

If the terminal moraines are overrun by succeeding glaciers a low spoon shaped series of hills will be formed. These are individually called drumlins and drumlin fields as a group.

Outwash takes the form of outwash plains when deposited on large flat areas and valley trains when deposited in confined valleys. Outwash differs from till in that outwash is sorted. These are excellent sources of sand and gravel since they require little sorting for commercial use.

Large masses of ice buried by outwash will eventually melt creating depressions in the outwash plain referred to as kettles. When filled with water they are kettle lakes, these abound in outwash plains.

Water running on the edges at the top of the ice will deposit loosely sorted materials on hillsides. These are kames and in groups they form kame terraces. Water running under the ice will make deposits of ridges of gravel and sand which snake across the countryside when the glacier has melted, these are eskers.

Where ice nears highlands there is water trapped between the two. These form lakes which endure as long as no lower outlet is formed. Many of these lakes have left deposits in the north country. Deposits in lakes by glaciers and the animal life and processes of the lake itself are glaciolacustrine.

Today about 8 per cent of the world is covered by active glaciers. At the height of the recent ice age about 30 per cent of the world was occupied by glaciers. Their deposits are found in mountain areas as well as flatlands. Northcentral United States and all of Canada have glacier deposits. Most of northern Europe as far south as Italy have glacial features. Most high mountain areas of the world today have glaciers, unique among these is the glacier on Mt. Kenya, a short distance from the equator.

Wind

Wind plays an important part in landscape formation in arid regions. However, it must be borne in mind that water, where it does occur in these regions, is still the dominant erosive force.

Wind erodes the land by deflation and abrasion. Deflated hollows in the ground are blowouts. If sand is channeled through the same area year after year small hills, yardangs, and troughs, yardang troughs, are formed. In wind eroded areas small stones called ventifacts which are faceted by wind abrasion are also found.

Pedestal rocks or teetering rocks were thought at one time to be caused by sand abrading the base of the formation. Today they are believed to be the result of differential weathering and erosion acting on alternate layers of weak and massive rocks.

Desert floors are usually covered with rock since most of the sand and dust is blown away. The stones are faceted and smoothed and are attacked by chemical weathering forming iron oxides, calcite, and silica. These rock floors are desert pavement.

When the material is blown away by the wind it is of course deposited somewhere else. Dust storms carry small particles high in the sky for many miles and when this is deposited it is loess. Loess covers large areas of Manchuria in Asia and in this country the area around Memphis, Tennessee.

Sand is piled up in dunes which have steep sides away from the wind and a gentle sloped side toward the wind. There are various kinds of dunes. Transverse dunes form at right angles to the wind. Longitudinal dunes are built parallel to the prevailing wind direction, when these develop slightly curved edges they are seif dunes. Curved dunes are barchans when the tines are pointed downwind and parabolic when the tines are pointed up wind. Different dune formations occur as a result of differences in vegetation and wind velocity. The highest dunes are the longitudinal found on the Libyan Desert, there they reach heights of 130 meters.

11

EARTH'S INTERNAL PROCESSES

There are several processes which operate within the earth, most of these are minor processes which do not fit into the scope of this work. The two dominant internal processes, the endogenic processes, are those of volcanism and diastrophism. Both processes tend to raise the earth's surface in elevation.

Earth's Internal Heat

Interior earth temperatures increase from the surface toward the center. These increases, of course, are not as great as those of the sun or stars. The increase in temperature with depth is referred to as the thermal gradient. In deep mines the thermal gradient has been measured at slightly over 25° C per kilometer. The value of the increase theoretically decreases in the earth's depths otherwise the heat at the center of the earth would be too high to be logical. The earth's core is believed to have a temperature of approximately 2500° C which is not high when the pressure at that location is considered. Molten lava of Mt. Vesuvius has been measured at 1210° C.

It is generally accepted that most of the earth's internal heat is the result of radioactivity near the surface. There is a relationship between silica content of rock and its radioactive heat producing capacity. The rocks deep within the earth are low in silica whereas near the surface such rocks as granite are high in silica. Therefore most of the earth heat produced by radioactivity is produced near the surface.

The internal heat of the earth escapes by volcanism and by conduction. It would be erroneous to think that this heat loss warms the atmosphere since the actual heat loss to the atmosphere is small and where volcanism occurs it dissipates very quickly. Heat energy can be transferred to other forms of energy such as mechanical energy which can be expended in uplifting mountain masses.

Volcanism

Volcanism is a term applied to the activity of molten earth matter. A volcano is a landform created by volcanism. Volcanic eruptions may be quiet or violent. Volcanoes created by violent eruptions may be classified into three broad groups, these are shield, composite, and cinder cone. Shield volcanoes have gentle slopes and cover large areas with their slopes seldom slanting more than eight degrees. The island of Hawaii is a shield volcano. Composite Cones resemble a child's image of volcanoes. These are dramatic peaks which loom to heights of 4000 meters possessing steep slopes and when dormant are usually snow capped. Mt. Fujiyama in Japan and Mt. Hood in the United States are examples of composite cones. The third category, cinder cones, look like large pimples on the landscape. They are rounded and seldom rise more than 500 meters and are usually found in clusters. Paricutin in Mexico is a classic example of a cinder cone.

Violent volcanoes have craters from which the materials are extruded and exploded. Craters are central pits which are deeper than they are broad. These may become calderas when the major central portion of the volcano collapses. Materials may also be exploded from rejuvenated caldera centers. Calderas are broader than they are deep. Crater Lake in Oregon is misnamed since it is a caldera and not a crater.

Volcanic materials may be gas, liquid, or solid. The most abundant gas of a volcano is water vapor. It is white and dramatic against the sky, but harmless. Lesser gases, more dangerous include carbon dioxide, carbon monoxide, hydrogen chloride, hydrogen sulfide, sulfur dioxide, sulfur, and ammonia. In some rare instances gases heavier than air are superheated and expulsed. These cause severe damage to villages in their path. The town of St. Pierre on Martinique was destroyed by such a gas from the volcanic mountain Pelee. Since then, this type of incandescent gas has become known as a Pelean Cloud, fiery cloud, or by the French name nuee ardente.

Liquid materials are termed lava and take many forms. Generally the smooth variety is called pahoehoe and the rough variety aa after Hawaiian designations.

Solid materials include the following: dust which is carried high into the sky and often around the world, ash which falls in the locality of the volcano, cinders or lapilli which are slightly larger than ash, bombs which are twisted cylinders of stone with hollow centers, and blocks which are angular. Bombs get their name from the explosion which results when the hollow centers expand upon heating.

Fissure eruptions are a type of nonviolent surface volcanism which extrudes a fine watery flow of lava usually composed of basalt. Many of the plateau areas of the world have been created by this form of eruption. The Columbia Plateau, Colorado Plateau, and the Deccan Plateau all display examples of fissure eruptions. Thus the deposits are called lava floods or plateau basalts.

When magma is formed deep underground there is usually pressure associated with it and the magma is forced into different underground rock areas. When these cool and are finally exposed by erosion or uplift the structures can be examined.

Underground igneous structures are plutons which may be massive or tabular. Tabular plutons are thin but cover a wide area, massive plutons are larger and have no definite shape ratio. If the magma forms sheets parallel to the strata of the native rock that it intrudes it is said to be a sill, if it cuts the strata it is a dike. Dikes and sills are tabular plutons.

Massive plutons include stocks which are large cylindrical intrusions less than forty miles in diameter, laccoliths which are dome shaped intrusions with flat bases, and batholiths which are huge deposits of igneous materials. Batholiths are usually found under folded mountain chains. Their long axes are parallel to the mountain axes. Batholiths are granitic in nature and some controversy exists over the origin of this granite. One theory is that granites are original intrusive volcanic rock while an opposing theory states that granites are the results of basic volcanic rocks melting the native rocks of an area and incorporating them into their structures. There is evidence that granites are created by both processes in nature.

Speculation regarding the origin of volcanoes is interesting. Measurement of volcanic explosions together with seismic records indicate that most volcanoes originate near the surface, that is, less than 60 kilometers deep. Original theories about the thermal gradient do not seem to account for enough heat to melt rock at this depth. Heat from radioactivity could melt rock but there is not much radioactivity connected with volcanic materials when they reach the surface. A more reasonable theory holds that the crust of the earth is shifting over the central portion and this shifting causes friction which melts the rock where slipping activity is at its greatest. There seems to be no doubt that the earth consists of plates which are moving away from each other on some parts of the earth and together in other places. For instance, the plates of the central Atlantic Ocean are moving away from each other. This could conceivably cause molten material to rise from this gap and create the oceanic mountain chain known as the Midatlantic Ridge.

Tectonic Processes

The uplifting of the earth has been divided into several categories each with its special terminology. The term diastrophism is used to designate a general process relating to the uplifting of a portion of the earth's surface. Tectonics refers to the movement of particular rock masses. Orogeny is the deformation of the earth crust in the development of mountains. Epeirogeny refers to the gradual elevation of large continental areas above sea level.

Two types of pressures or forces are found within the earth, compressional and tensional. Compressional forces move toward a point and tensional forces move away from a point. These forces cause rocks to bend, break, and move. Deeply buried rocks will become plastic and bend under pressure but surface or near surface rocks will become plastic until they reach their stress limit and then they will break.

Bent rocks are said to be folded. These are formed by compressional forces into two general shapes. The two major folds are anticlines and synclines with variations of these further identified for exact scientific purposes. Anticlines are uparched folds while synclines are downarched folds. They almost always occur together. A simple uparched structure with strata sloping away from the center is a dome. The Ozark Mountains are an example of a domal structure.

Rocks which break along the horizontal strata are said to exhibit sheeting. When cracks occur vertically they are joints

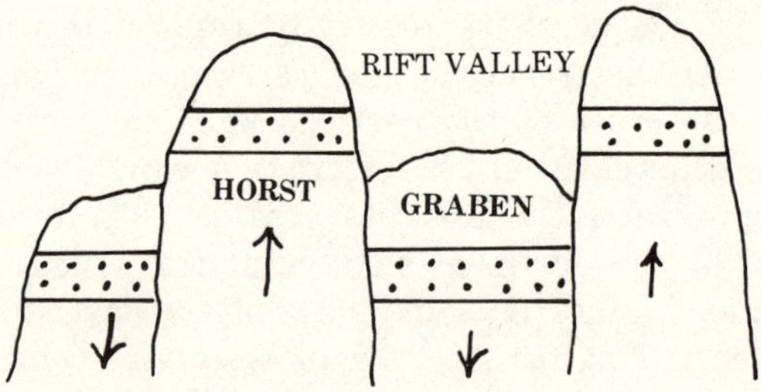

Figure 5 a. Block faulting occurs when masses of earth materials move up and down along vertical fractures.

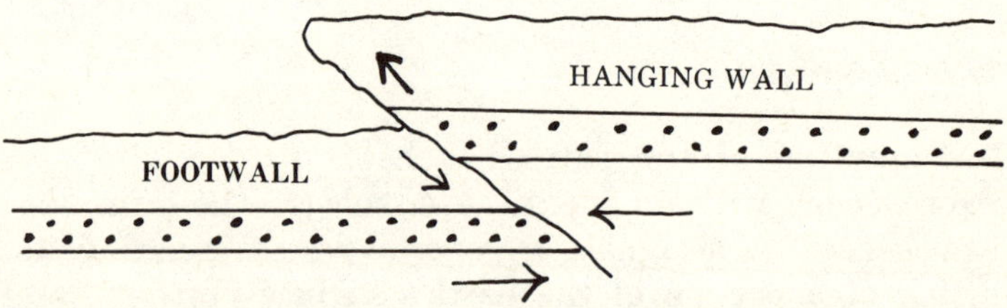

Figure 5 b. A reverse or thrust fault occurs when the hanging wall is forced up and over the foot wall. This is the result of the action of compressional forces.

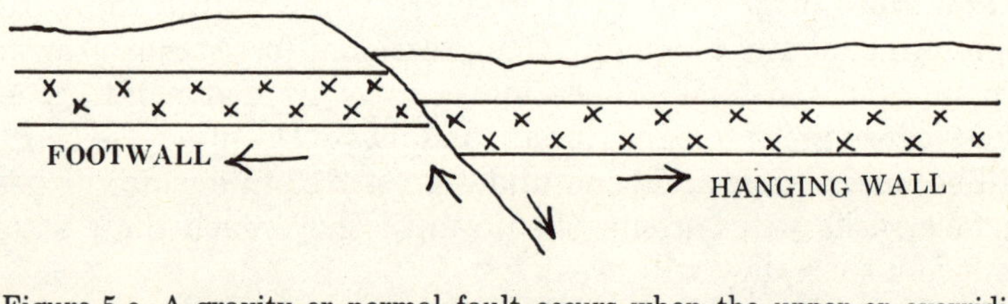

Figure 5 c. A gravity or normal fault occurs when the upper or overriding block moves downward in relationship to the lower block. This is the result of tensional forces acting away from a point.

Figure 5. Three types of faults

and when movement occurs along any crack these are termed faults. Faulting can be the result of either tensional or compressional forces.

If the faulting is vertical then one block of material moves down and one moves up in relation to the other. The upraised block is a horst and the sunken block is a graben. Refer to figure 5 a. These are block faults and the resulting depression is a rift valley. Such valleys make up central Pennsylvania. In Africa, a rift valley extends from Madagascar up through lakes Tanganika and Nyassa to the Red Sea and finally to the Adriatic Sea and to the base of the Alps.

If the faulting is at an angle then one block rides upon the other. The upper block is the hanging wall and the lower block is the footwall. Compressional forces acting on a fault causes the hanging wall to move upward and over the footwall. These faults are reverse faults or thrust faults. Refer to figure 5 b. On the other hand, tensional forces cause the upper block to fall along the fault and the lower block to be higher in elevation. These are normal or gravity faults. See figure 5 c. Faults which move diagonally, that is, both horizontally and vertically are oblique faults. Faults where the movement is horizontal are strike-slip faults. The term strike refers to the horizontal direction which is found at right angles to the direction in which the fault slants or dips. The high walls exposed by faults are fault scarps and streams are often found to follow these fault lines.

Earthquakes

When faults occur, the breaking rock sends a shock wave to other parts of the earth. It is as if a rubber band were stretched between two hands. As the band reaches the elastic limit it begins to quiver until it breaks and then the parts slap back and forth at the hands holding them. Rocks react in much the same manner, they send out small waves called foreshocks before the break, then there are the waves of the break or main shock and then the waves of recovery called aftershocks.

The actual point of the earthquake is the focus and the location above ground is the epicenter. Most of the earthquakes occur within 30 to 80 kilometers of the earth's surface.

Earthquake waves are seismic waves and they are measured by an instrument called a seismograph. Seismographs are set in bedrock and a shock somewhere on earth is transmitted through this bedrock to all locations on earth. Shocks are graded on a basis of 10 with 10 being very severe. Shocks of 7 can cause great damage to inadequate housing. Generally shocks of 3 and 4 are felt by people but do not cause damage.

There are about a million earthquakes a year but only a few are of sufficient destructive force to make the news. About 80 per cent of all earthquake energy is centered around the western Pacific Ocean, another 15 per cent around the Caribbean, and about 5 percent around the Mediterranean.

The waves sent out by earthquakes are of two main types, body waves and surface waves. The surface waves have little scientific value but the body waves indicate the internal structure of the earth. Body waves are primary and secondary. Primary waves or P waves are push-pull waves. These are longitudinal waves which compress materials similar to one's voice compressing the air in front of the mouth. Primary waves can travel through liquids, solids, and gases. Secondary waves or S waves are shake waves. These are transverse waves which move similar to waves on water or to waves created when a rope attached on one end is shaken up and down. Secondary waves only travel through solids. If an earthquake occurs on one side of the earth all recording stations should receive both P and S waves. However, stations to the sides and directly opposite on the earth only receive P waves. Hence it has been concluded that the center of the earth is composed of heavy metals in a plastic state.

The Earth's Interior

On the basis of seismic measurements the earth's interior is usually divided into three main parts, the crust, the mantle, and the core. Further divisions have been made, the most significant of these is that of dividing the core into inner and outer areas. See Figure 6.

The crust of the earth is an outer shell which varies in thickness from 16 kilometers under the oceans to a continental thickness of 64 kilometers. The upper area of the crust is light rock and the lower portion of the crust is heavier. The lower crustal limits were identified by the Yugoslavian scientist Mohorovicic who studied hundreds of seismic records. This lower boundary was named the Mohorovicic Discontinuity in his honor, now called moho for short. Attempts at reaching this level by drilling the mohole have not yet been successful.

The mantle which lies just below the crust is roughly 2880 kilometers thick. Since P and S waves travel through it at high

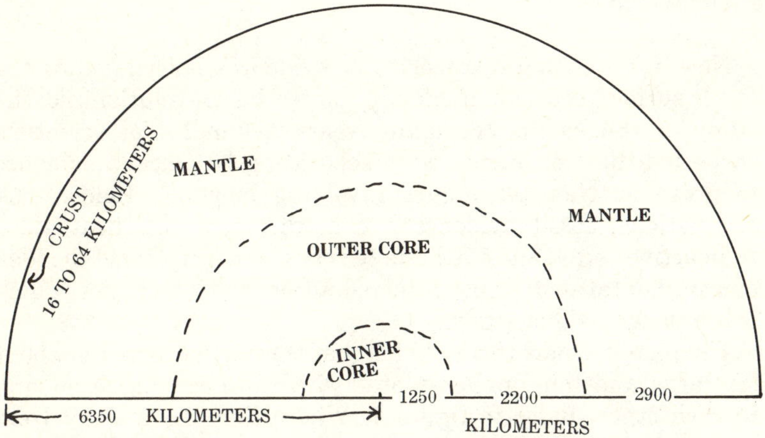

Figure 6. Divisions of the earths interior with approximate thicknesses.

speeds it is believed to be a dense solid. Rocks composing the mantle are probably high in iron and magnesium associated with silica. These rocks could be variations of ultrabasic igneous rocks called gabbro and dunite.

The central portion of the earth is a core with a radius of 3456 kilometers. It is believed to be a nickel-iron alloy obviously under high pressure. This confining pressure causes the metal to behave as a plastic. The inner section which is about 2496 kilometers in diameter transmits P waves differently than its outer shell and on the basis of these differences is believed to be solid. No explanations seem reasonable for the differences of the inner and outer cores.

The reason that the central core is believed to be an iron-nickel alloy is that the earthquake wave transmission of the core corresponds well to that of these combinations at the surface. Although other heavier metals could be at the earth's center they are rare and are thus ruled out. One method of possibly deciphering the earth's composition is by the study of meteorites. Meteorites are believed to be typical of solid materials of the solar system. They are of two basic types, stoney meteorites and metallic meteorites. The metallic meteorites are almost all iron-nickel combinations, thus supporting the theory of the earth core composition.

Plate Tectonics

New and exciting discoveries of evidences showing that the earth surface consists of moving plates has revolutionized the study of the earth. For many years a handful of scientists suspected that continents were separating but lacked evidence to prove it. New techniques involving magnetic fields, rock correlations, fossil evidences, coral displacement, heat flow, radioactive decay, and earthquake zones have reinforced the idea of continental drift, now referred to as plate tectonics which is a broader field of study.

The plate tectonic theory holds that the earth consists of about six large and ten smaller plates which are moving from and to each other. Refer to figure 7. Where they move away from each other block faulting and volcanism occurs. These are expressed in the ocean as ridges where volcanism occurs, and rift zones where volcanism is lacking. The major ridges occur in the middle of the Atlantic Ocean, the Indian Ocean, and in the south Pacific Ocean extending in an arc to the east of South America. Where the plates collide there are thrust faults and volcanic action. These are the most intense volcanic and earthquake zones of the world. The major areas of collision are at the western Pacific Ocean rim, from Central America south along the South American coast, and in a line from the Adriatic Sea to Pakistan.

Figure 8 is a diagramatic sketch at about 30° South Latitude showing the western Atlantic or American Plate moving westward and colliding with a small plate of the eastern Pacific. Perhaps this collision has caused the Andes Mountains to rise. In the middle Atlantic the African plate moves away from the American plate allowing magma and lava to escape creating the Midatlantic Ridge. Measurements of past magnetic fields and radioactive dating of the lava indicate that locations furthest from the ridge have the oldest date and the materials become younger toward the ridge. This indicates a gradual spreading of the sea floor which can be measured. The rate of spreading is about 6 to 9 centimeters per year.

Where the plates collide the overriding plate may become an arc shaped island chain and the underriding plate an ocean trough or trench. Such trenches and island arcs abound in the western Pacific Ocean.

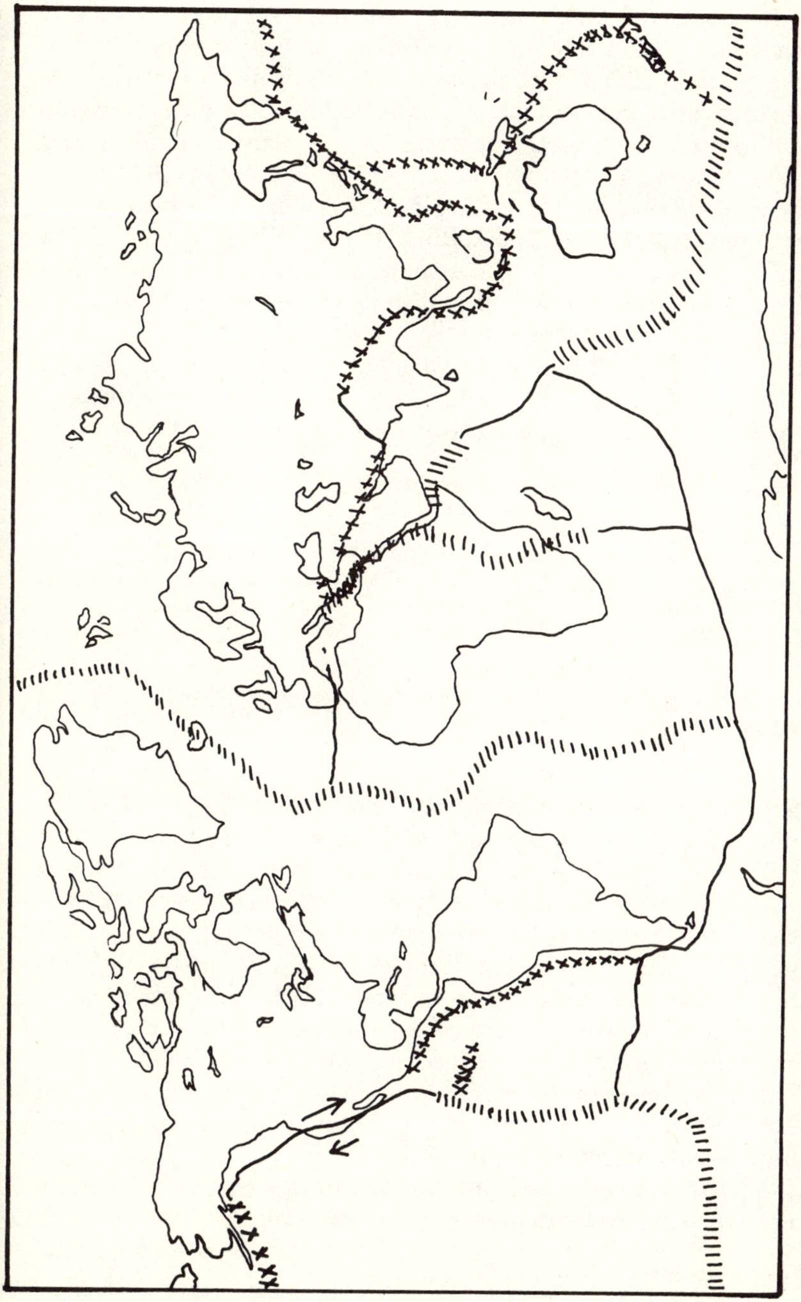

Figure 7: Simplified sketch map depicting major and some minor crustal plates of the world. (=) indicates spreading plates (xx) indicates plates coming together, arrows indicate shearing plates

It is believed that the crustal plates have broken apart and have been set in motion by convection or heat currents within the mantle. Even though deep measurements of convection currents have not been made the surface heat loss measurements indicate that hot earth materials of the mantle circulate and cool much as air in a room circulates and cools. Heated air in a room rises to the ceiling and gives up some heat which is conducted through the ceiling. If the ceiling has separated in some spot then a larger amount of heat escapes there. The heated air in the room cools and sinks to the floor to be replaced at the ceiling by warmer air. So it may be with the heat within the mantle. Heat in the earth mantle rises to the crust and moves along it and slowly through it by conduction. Where a weak spot occurs in the crust heat escapes faster. This is borne

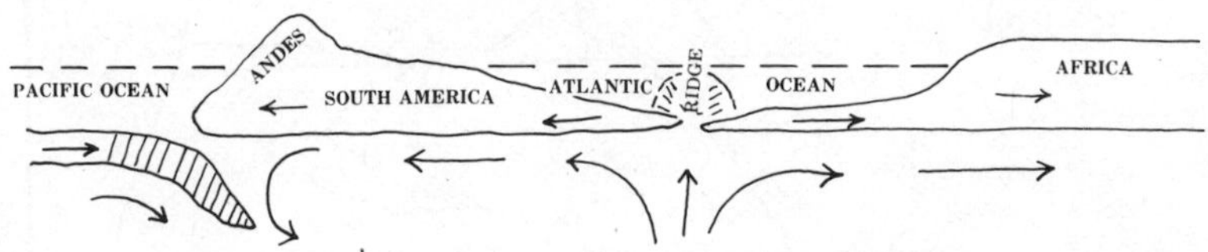

Figure 8: Sketch about 30° S depicting sea floor spreading and underground thermal currents.

out since the highest heat loss from within the earth occurs where the plates are separating at the oceanic ridges.

The process which sets the convection currents in motion is still a mystery but there are logical explanations which may be correct. One possible explanation is that the heat is built up within the earth by the decay of radioactive materials. This heat causes the materials to expand and the direction of least pressure is toward the surface. The heated materials then move very slowly toward the surface, only a few centimeters a year but in a convectional movement. As the materials reach the surface they move along the Mohorovicic Discontinuity and occasionally break through it. They cool and become denser under the crust and sink back into the mantle interior to be replaced near the surface by lighter warmer materials.

Stable Continental Areas

Continents and continental areas can be divided into several groups depending upon geologic age and rock formations. For instance, North America can be divided into four regions. It has a stable ancient rock interior exposed in central Canada and in the United States around Lake Superior and the Adirondacks. This is the Canadian Shield. The second area would be central United States where horizontal strata overlies the buried shield. The third area could be the folded mountains flanking the interior platform and the fourth area could be the sediment accumulating basins off the coastal areas.

The stable shield areas are the oldest dated rock areas of the world. Evidence indicates that they have never been under deep water and seem to be gradually wearing down. These areas are believed to be the ancient exposed land of the earth which was crowded into one small area and which has broken apart and presently make up the core of the shifting plate areas. The shield areas include the Canadian Shield of North America, the Guiana Highlands and southern Brazil in South America, Scandinavia in Europe, Siberia and central India in Asia, most of Australia, and central Africa. It is onto the edges of these stable areas the sediments are welded and the continents expand. When the sediments are folded the continents are once again shortened. The new highlands are then eroded and deposited back into the newly created basins.

12

SUMMARY OF ENVIRONMENTS AND PROCESSES

The environments of the earth's surface are subject to the processes operating upon the earth. These environments are creations of the atmosphere and the agents operating at the surface. Major land environments include mountains, plateaus, hills, and plains, each having their own types of minor variations, usually in the form of valleys. Each is capable of having wind, water, and ice working upon it.

More specific classification of environments may be made under three broad categories; these are (1) continental, (2) shore, and (3) marine.

The continental environments include those of the desert and the ice fields. There are also fluvial environments of the continents which are found in the mountain highlands, the piedmont hills, and the valley flats of the lowlands. There are other continental environments of water which include the lacustrine or lake, the spelean or cave, and paludal or swamp. The shore environments include the beach, the delta, the lagoon, the estuary, and the littoral which is the tide zone. The marine environments include the open ocean, the neritic or shelf zone, the bathyal or slope, and the abyssal or ocean deep. Figure 9 is a general sketch of the relationships of these environments.

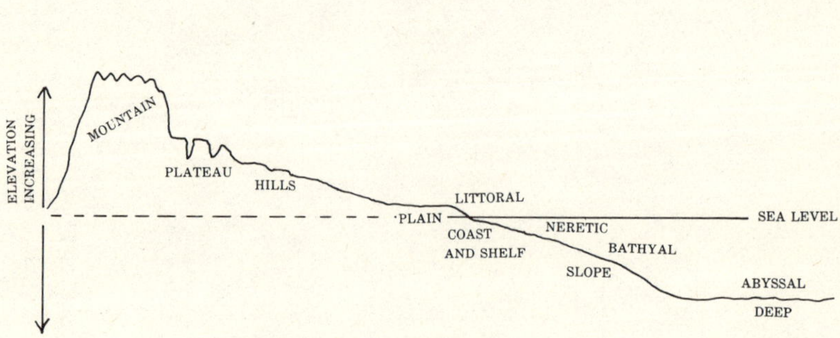

Figure 9: Basic earth environments

Climates

The climates include the tropical designations of rainforest, savanna, and monsoon, the dry climates of desert and steppes, the middle climates of humid subtropical, mediterranean, marine west coast, and midlatitude monsoon, the snow forest climates of humid continental and subarctic, and the frost climates of tundra and ice cap. The relationships between climates are summarized in figure 10.

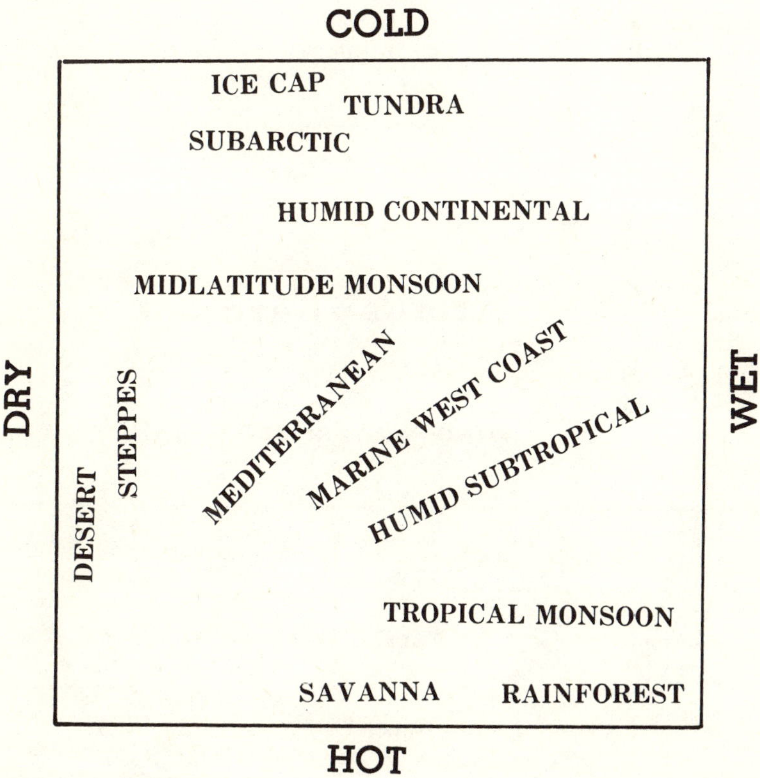

Figure 10. Climates. The chart illustrates the relationship of the major climates as to temperature and precipitation. Two scales exist. The horizontal scale illustrates precipitation extending from dry on the left to wet on the right. The vertical scale illustrates temperature, extending from hot on the bottom to cold at the top. Climates are placed in a comparative position in reference to their attributes of temperature and precipitation.

Earth Processes

The energy of the earth flows in an endless stream of processes from one form to another. It moves in rhythms and cycles. The major earth processes are graphically illustrated by figure 11.

On the earth surface the materials are subjected to weathering by the elements of the atmosphere, the weathered materials are eroded, transported, and deposited as sediments by wind, water, and ice. The sediments become rock and part of the lithosphere. These are brought further underground by more deposition and are subjected to the internal processes of volcanism and diastrophism.

Materials of the earth are influenced by the rotation of the

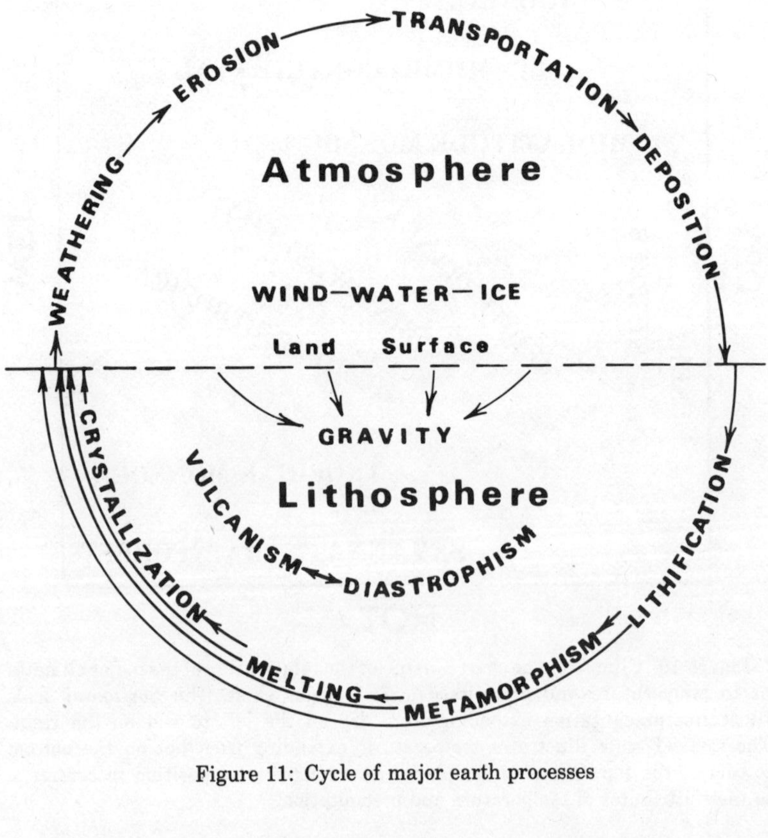

Figure 11: Cycle of major earth processes

earth, gravity, magnetism, and the energy supplied by the sun. The earth is constantly supplied with new energy from the sun and it in turn expends or stores this energy. The earth is only one small unit in a universe of energy.

INDEX